CANADIANS
IN SPACE

CANADIANS IN SPACE

The Forever Frontier

JOHN MELADY
Foreword by Henry Champ

DUNDURN PRESS
TORONTO

Project Editor: Michael Carroll
Copy Editor: Cheryl Hawley
Designer: Jennifer Scott
Printer: Webcom

Library and Archives Canada Cataloguing in Publication

Melady, John
    Canadians in space : the forever frontier / by John Melady.

ISBN 978-1-55002-940-6

    1. Astronauts--Canada--Biography. 2. Outer space--Exploration--Canada--History. I. Title.

TL789.8.C3M45 2009   629.450092   C2009-901987-63

1   2   3   4   5   13   12   11   10   09

We acknowledge the support of the Canada Council for the Arts and the Ontario Arts Council for our publishing program. We also acknowledge the financial support of the Government of Canada through the Book Publishing Industry Development Program and The Association for the Export of Canadian Books, and the Government of Ontario through the Ontario Book Publishers Tax Credit program, and the Ontario Media Development Corporation.

Printed and bound in Canada.
www.dundurn.com

Dundurn Press	Gazelle Book Services Limited	Dundurn Press
3 Church Street, Suite 500	White Cross Mills	2250 Military Road
Toronto, Ontario, Canada	High Town, Lancaster, England	Tonawanda, NY
M5E 1M2	LA1 4XS	U.S.A. 14150

*For Maureen Paulhus,
with admiration and thanks
from your old godfather.*

Canadians Who Have Been in Space

Roberta Bondar: Born December 4, 1945, Sault Ste. Marie, Ontario. Mission Date: STS-42, space shuttle *Discovery*, January 22–30, 1992.

Marc Garneau: Born February 23, 1949, Quebec City, Quebec. Mission Dates: STS-41-G, space shuttle *Challenger*, October 5–13, 1984; STS-77, space shuttle *Endeavour*, May 19–29, 1996; STS-97, space shuttle *Endeavour*, November 30 to December 11, 2000.

Chris Hadfield: Born August 29, 1959, Sarnia, Ontario. Mission Dates: STS-74, space shuttle *Atlantis*, November 12–20, 1995; STS-100, space shuttle *Endeavour*, April 19 to May 1, 2001.

Steve MacLean: Born December 14, 1954, Ottawa, Ontario. Mission Dates: STS-52, space shuttle *Columbia*, October 22 to November 1, 1992; STS-115, space shuttle *Atlantis*, September 9–21, 2006.

Julie Payette: Born October 20, 1963, Montreal, Quebec. Mission Dates: STS-96, space shuttle *Discovery*, May 27 to June 6, 1999; STS-127, space shuttle *Endeavour*, July 15–31, 2009.

Robert Thirsk: Born August 17, 1953, New Westminster, British Columbia. Mission Dates: STS-78, space shuttle *Columbia*, June 20 to July 7, 1996; Expedition 20/21, Soyuz TMA-15, May 27 to November 2009.

Bjarni Tryggvason: Born September 21, 1945, Reykjavik, Iceland. Mission Date: STS-85, space shuttle *Discovery*, August 7–19, 1997.

Dave Williams: Born May 16, 1954, Saskatoon, Saskatchewan. Mission Dates: STS-90, space shuttle *Columbia*, April 17 to May 3, 1998; STS-118, space shuttle *Endeavour*, August 8–21, 2007.

CONTENTS

FOREWORD

This book is a delight and a must-read for those millions of Canadians, and millions is the right number, who take pride in Canada's contributions in outer space.

A big part of John Melady's effort centres on the Canadian Astronaut Corps, bringing both a sharp image of who they are at work, and who they are privately. Why they commit many years of their lives to what is clearly a great adventure, but in reality is such a short time space.

Melady's narrative fills two other very important gaps: a brief but necessary review of the history of space exploration, and the work of early Canadians who saw the stars and wanted to know more — all the while dreaming of what could be.

There is much about the technology that Canadian industries have made: the Canadarm, which has led Canada to a world-lead in robotics; Canada's gains with space cameras and imaging. Things that often go unnoticed by those of us who are mesmerized by the flight itself, the majestic takeoffs and the tranquil landings.

There are also descriptions of failures, missed opportunities, and concerns for the future. In short, this book serves as a dictionary, an encyclopedia, and the bible of Canadians in Space.

On a personal note, I have had the opportunity to cover space exploration since the sixties, when I would cover the overnight and very early morning broadcasts on Gemini and Apollo missions, sitting in for Harvey Kirck, who was CTV's space guru. Fortunately, I was involved at both NBC and CBC covering launches, in-space events, and landings. I, like everyone else who witnesses a liftoff, can never cease marvelling at these events.

I've met John at the Cape. I admire his tenacity and professionalism, in short, the way he does his work. This book is a tribute to his craft and a boon for his readers.

HENRY CHAMP
SENIOR CORRESPONDENT
CBC NEWSWORLD
WASHINGTON, D.C.

ACKNOWLEDGEMENTS

The research and writing of this book took three years, but it seemed like five. At times, in fact, I wondered if there ever would be an end in sight. There were several reasons for this, almost all of them beyond my control. First of all, I did not feel I could do the book without interviewing the brave men and women who are our astronauts. Despite the fact that they have received much media attention over the years, a quick perusal of a few newspaper stories is not the same as meeting face to face. Fortunately, in large measure, I was able to do so.

I also knew that I needed to spend time at the Canadian Space Agency in Montreal, the Kennedy Space Center in Florida, and the Johnson Space Center in Houston. I went to all — and because I was determined to observe at least one shuttle launch, I ended up going to Kennedy four times: twice during specific research trips for this book, and a couple of times before that. At long last, the weather, the mechanical factors, and the scores of other problems that delay launches were not present. I finally saw a shuttle fly.

Contacting, and then finding, mutually convenient interview times with astronauts was never easy. They keep hectic schedules involving training, space missions, international travel, public relations duties, and, whenever possible, snatches of personal time. I did not want to intrude, and for the most part, I don't think I did. In every case, when we finally found time to sit down and let the tape recorder run, I found Canada's astronauts to be as hospitable and helpful as any of the hundreds of individuals I have interviewed over many years of writing. For that reason, I thank them profusely for all their assistance on this project. They are indeed, admirable individuals.

As always, there are many others who helped me in a myriad of ways. I am truly indebted to each, and I want them to know that their efforts helped make this book. In Montreal, Media Relations and Information Services expert Jean-Pierre Arseneault went out of his way to assist. He arranged interviews, spent time following-up on my requests, arranged for me to see all I needed to see at our Space Agency, and made suggestions that were pointed and positive. Others at the Space Agency assisted as well: among them Carole Duval, Julie Simard, Catherine Vallée, Magalie Renaud, and Benoît Marcotte.

At Cape Canaveral, Margaret Persinger in Kennedy Space Center Public Affairs was always cheerful and helpful in suggesting and then locating photographs that I needed. Laurel Lichtenberger was especially kind to me as I jumped through the various hoops in obtaining media accreditation. Bert Ulrich, Multimedia Manager for NASA in Washington, cut through red tape and made it possible for me to obtain necessary documentation in a timely manner. Many Virata was always gracious and knowledgeable, and never lost his composure when responding to my questions. Jean-Louis Santini, Science Correspondent with Agence France-Presse, and Sidhartha Banerjee, of Canadian Press, were helpful desk mates at KSC.

My work in Houston and at the Johnson Space Center was made much easier because of the following people: Gayle Frere, John I. Petty, and Aaisha Ali. Mark Carreau, senior space reporter for the *Houston Chronicle*, provided insights into how things worked at JSC. But it was Kathy and Frank Koeck who made my time in Texas particularly memorable and productive. I owe them a special debt of gratitude.

In Ontario, Tania Alves was instrumental in setting up interview time with a very busy Doctor Dave Williams, and her doing so was much appreciated. Bill McMaster and his colleague John Nowlan provided helpful assistance. In the latter case, the KSC information and photographs that John passed along were what I needed at just the right time. My conversations with Brian Butters about NASA and its intricacies were helpful, as was the information provided by Doctor Tom Drake and Ken Papple. Carolyn Gossage acted as a conduit to Roberta Bondar, and that was appreciated. Michele Melady, Manager of the CBC Reference Library in Toronto, made a family connection productive, while Crystal Tupling and David Jaques, in London, were instrumental in so much of the photo work that I needed for the book.

At Dundurn, Beth Bruder, Ali Pennels, Barry Jowett, Michael Carroll, Jennifer Scott, Margaret Bryant, and my editor Cheryl Hawley were professional and personable in all their dealings with me. But I owe two other individuals at Dundurn a special thank you: Tony Hawke was there during all the stages of the project, from contract preparation to the final edit. Kirk Howard, whose decision it was to publish this book, was the key factor in its coming to fruition. Anyone reading these notes should bear in mind that if Kirk had not been there, neither would this book.

Henry Champ, Senior Correspondent for CBC Newsworld in Washington, was somehow able to find the time to write the foreword to this volume. Henry is a true professional in a hectic profession, and I am extremely grateful for his insight, advice — and yes, for his words as included here.

Closer to home, bookseller Tom Fincher in Goderich, Ontario, was always supportive, while librarians Anne Dodington and Jeanette Finnigan never complained when I used the interlibrary loan service as if it was my own. And lastly, but to me the most important person of all, is my wife Mary, who put up with me while this piece of research and writing was being done. I am now more indebted to her than ever.

John Melady
Seaforth, Ontario
July 1, 2009

I

A Pillar of Fire, Blinding and White

S hortly before the launch of the spaceship *Endeavour* from Cape Canaveral, Florida, on Wednesday, August 8, 2007, spectators were told what
to do if the thing blew up. "Go indoors immediately," warned the voice
over the public address system. "If you are in your car, keep the doors and
windows closed."

Because of these few words, images of the catastrophic explosion of
Challenger came to mind. That was the space shuttle that disintegrated
above the heads of thousands of horrified spectators on January 28, 1986,
and was seen by millions more on television. On that terrible day, seven
crewmembers died instantly, one of whom was Christa McAuliffe, a young
mother of two from a place called Concord, New Hampshire. She was a
teacher who dreamed of conducting classes in orbit. But that was not to be.

As the throng of shocked observers stared skyward, an undetected
mechanical problem caused the gleaming white spaceship to tear itself
apart, snuffing out the lives of all on board. The flight lasted a mere
seventy-three seconds.

And now, with *Endeavour* ready to go, concerned officials at the Kennedy Space Center warned those closest to the launch pad to use caution as the countdown continued. Noxious gases and debris could blanket the nearest viewing areas if there was an accident. Obviously, the National Aeronautics and Space Administration, NASA, did not want those who might bear witness to tragedy to be hurt themselves.

I was at the desk assigned to me in the media compound at Kennedy when the sobering message was broadcast. As the words of warning hung in the air, I glanced at those closest to me, to watch their reaction. In essence, there was none; but it was not that the announcement had not been heard. An arm's length away, Carole Duval, of the Canadian Space Agency's Montreal office, paused momentarily as she spoke to colleague Jean-Pierre Arsenault, but then quickly continued with what she had been saying. Across the room, neither veteran space reporter Mark Carreau, of the *Houston Chronicle*, nor Sidhartha Banerjee, of the Canadian Press, looked up from their laptops. It was left to the insightful newsman who shared my workspace to put the P.A. message into context for me:"NASA

The central media room at the Kennedy Space Center at Cape Canaveral in Florida. Here, reporters from all over the world work when they are covering space shuttle launches.

is covering its ass," muttered Jean-Louis Santini, science correspondent in North America for Agence France-Presse. He was right of course.

Meanwhile, as the launch countdown moved inexorably to zero, the tension, even among seasoned news people, moved in the opposite direction. Finally, just before the countdown was about to end, most of those present in the media building left their unfinished dispatches and surged outside to watch the spectacle. Hours, even days earlier, remote cameras with long lenses had been put in place to film the thing. Despite a twenty-four hour hold earlier on, this launch would now go on time. Nevertheless, nothing about the whole undertaking could be taken for granted.

I went outside with all the rest, and while I cannot speak for anyone else who stood transfixed watching Pad 39A and the shuttle that stood upright on it, I felt that my emotions were profoundly conflicted. While I wanted — we all wanted — this launch to be successful and safe, there was still a palpable sense of danger in those final critical minutes.

I was only at Kennedy for one launch, mainly because an earlier one I waited for was postponed for two weeks. Staying in a Florida hotel and having to endure several days of rain is not particularly pleasant, so on that occasion I caught a flight home instead. The mission I missed did give me a chance to familiarize myself with KSC protocol, accreditation, and personnel, but not a launch. That time though, and again now, things could go wrong.

That feeling is not uncommon. Jay Barbree, one of the veteran reporters at space launches, who covered so many of them for NBC television, told me that "there is always tension in the air, particularly in the last stages of the countdown."[1]

In fact, Barbree mentioned the same sense of apprehension in his recent book about the American space program: "No matter how many of these shattering launches you have seen," he writes, "No matter how many times you have felt the body-shaking impact, the shock waves rippling your clothes and skin, you never feel at ease."[2]

And now, as reporters and others fanned out across the three acres or so in front of the press buildings, the big television trucks stood stately in a nearby parking lot. On the flat tops of their own buildings, the network cameras were good to go. The broadcasts would be seen live on countless televisions across the country and around the world.

The shuttle on 39A was roughly three miles away, but spectators were not allowed to venture any closer to it. After all, if the rocket exploded on site, a vast area around it would have been levelled, and anyone too close would die. Some said that only the after-effects of an atomic explosion would be comparable in scope and devastation caused. There was a precedent for that.

The space shuttle *Endeavour* sits on Launch Pad 39A at Kennedy Space Center in Florida, on August 5, 2007. Three days later, Dave Williams and his crewmates left for the International Space Station. The spelling of the name of the shuttle annoyed some Americans.

Some thirty-eight years earlier, on July 3, 1969, a large Soviet vehicle had exploded on its pad. The accident happened at the grim and, at the time, secret Baikonur Cosmodrome in Kazakhstan. That day, the unmanned rocket being tested blew apart, and not only was the machine and the pad it rested on obliterated, but so was every bird and beast in a wide swath on the surrounding steppe. The explosion was believed to have had "a strength equivalent to 250 tons of TNT, not quite the power of a nuclear explosion, but still formidable." A month later, through the American photoreconnaissance spy program, "the ruin at Baikonur"[3] was clearly identified. Amazingly, no human being was hurt in the incident.

So, here at the Kennedy Center media site we were not the only ones kept well back from Pad 39A. The thousands of vacationers, residents, and the merely curious who flocked to see this launch were also constrained in where they could go. Local media identified the restricted areas and even ran maps in the papers that outlined the boundaries. These maps not only referred to off-limits areas on land; equally extensive prohibited watercourses, lakes, and even areas of the adjacent Atlantic Ocean were included. The newspapers played it both ways though — the so-called "best viewing areas" were sketched out as well.

The last minute of the countdown was tension-filled, spellbinding, and breathtaking. We all stood in relative silence, a silence that was a mixture of anticipation and perhaps even dread. Our eyes never left the shuttle: solitary, motionless, clearly highlighted above the scrub trees, gleaming against the backdrop of the early evening sky. In front of the crowd was the large, digital countdown clock that is almost as old as the space program itself. This is the same clock that was there during the Apollo missions to the moon. We have all seen it, time and time again, in the televised pictures of space shots from the past.

The public address announcer cut in as the time on the clock reached minus ten. What followed had all the elements of drama on centre stage: the main event, the enraptured audience, the denouement; swift, dramatic, unforgettable.

"Ten, nine, eight, seven, six — go for main engines start — four, three, two, one, zero — and liftoff of space shuttle *Endeavour!*" The words came with a flourish, like an arena announcer's dramatic introduction of the home team starting lineup at a sporting event. At this place though, there was far more at stake than at any game, anywhere. There were lives on the line; the lives of seven brave men and women who were doing what few others would ever dare to do.

Initially, there was no visible movement, nor was there sound. A couple of seconds later, however, that all changed. Flame from the three engines of the shuttle and the two solid rocket boosters lit the hazy sky.

The rocket began to rise. Then the thousands of gallons of water that are used for sound and flame suppression at the pad turned into towering clouds of boiling steam in the white-hot heat of blastoff. The upward climb of the shuttle seemed slow at first, largely because of the enormous forces needed to hurl it to the heavens. In no time, the *Endeavour* was higher than the trees and the great pillar of fire below it was blinding and white.

That's when the sound hit.

With a roar like thunder, and a jolt like a powerful punch to the chest, the massive wave of sound clobbers you, takes your breath away, weakens your knees, and shakes the soil on which you stand. This is an awesome spectacle that you will never, ever forget. In that moment, I saw tears. There were also screams, laughter, cheers, and even prayers. The man at my left crossed himself, and then as if embarrassed, put a rosary back in his pocket. I found myself hooting with excitement, even as I tried to take pictures, record the sound, and observe the hurtling spaceship and those who watched it.

My feeling was shared by everyone there, and there have been similar reactions at every launch. Years ago, CBC television anchor man Knowlton Nash remembered how he felt every time he saw what I had just seen: "Each rocket would rise on liftoff with such agonizing slowness; you'd find yourself screaming at the thing to get off the ground. The screams would turn into cheers at it rose majestically over the beach and headed into the sky over the Atlantic Ocean. Then suddenly, you'd fall silent, embarrassed at the yelling but realizing that everybody else had been yelling, too."[4] To me, only the half-forgotten memory of a volcanic eruption carried more drama.

Less than ten seconds after liftoff, the first calm, low-key, laconic transmission from the flight deck of the *Endeavour* echoed from the P.A. speakers. An immediate reply came, not from the launch firing room a couple of hundred yards to my left, but from the Mission Control room in a large, nondescript, windowless building, one time zone and hundreds of miles away. Henceforth, all communication to and from this flight would be through Houston, Texas. The launch alone is Kennedy-controlled, and even though shuttles leave from here, and if possible land here, it is Mission Control at the Johnson Space Center in Houston that is the nerve centre for this and all manned missions.

In front of us, the *Endeavour* streaked into the sky, as the ever-increasing staccato-like rolling roar of its engines obliterated even the cheers. The long, white, billowing plume of exhaust grew more pronounced as the rocket gained altitude and speed. Conversely, the shuttle itself grew smaller until it was but a glowing white star, hurtling out of sight and into space. Five minutes after the launch all that remained were the wind-tossed exhaust clouds and the ever-softer echo of the sound. A short time later, even they were gone.

Mission Control announcer Rob Navias, in Houston, broke into the reverie of the moment when he began identifying the astronauts who were now far above us, still strapped in their seats and marvelling at where they were. Commander Scott Kelly was named first, then his compatriots. Two of the crew who occupied middeck seats were of particular interest to me,

Courtesy John Melady

Space shuttle launches are always dramatic. The white-hot flames from the great engines can be seen for miles, and the roar produced is ear-splitting. Here, spectators watch the early evening departure of *Endeavour* on August 8, 2007. Canadian astronaut Dave Williams was on board.

and I found myself listening intently for their names. One was a fifty-five-year-old woman from Fresno, California. The other was two years younger, a man born in Saskatoon, Saskatchewan. This was the first flight for the American; the second for the Canadian. She was a teacher; he a doctor.

The teacher's name was Barbara Morgan, and she was making this flight to prove a point. Years earlier, she had been among the many who stood and watched as her friend Christa McAuliffe had gone to her death on the ill-fated *Challenger*. Back then, Morgan was McAuliffe's backup, and knew that had Christa been unable to fly that day she, Morgan, would have died instead. But today she was going into space herself. She was even sitting in the middle seat on the middeck, as McAuliffe had.

Morgan's mission had been highly anticipated. In the papers, on television and radio, and in countless internet references, she would fulfill the stated purpose of many: to put an educator on a shuttle. The idea had been long in coming to fruition, starting as far back as 1984, when then U.S. President Ronald Reagan announced his Teacher in Space program. That was when Barbara Morgan responded to the challenge. She, and more than eleven thousand others, applied to fill the role, but only she and McAuliffe were selected. However, over the years the requirements for the job changed.

By 1998, NASA still wanted a teacher to fly, but now the person who would do so had to be a qualified astronaut as well. Morgan was still interested, and after her basic training had been completed she obtained further experience in shuttle communication and robotics. By the time she was named to a crew, she had much more than her teaching background and certification alone. She was now a fully trained Mission Specialist. She was also, as NASA Administrator Michael Griffin told the press, not only "very likeable," but "tough as nails underneath."[5] She was truly the ideal replacement for McAuliffe.

The doctor on the shuttle was Dave Williams, a career astronaut whose inclusion on the flight had brought me to Cape Canaveral in the first place. His background, education, and career will be detailed later on, but suffice to say, on this flight, as on an earlier one, he performed admirably. He is one of the best, and I felt so privileged to be there to witness this launch of a countryman. Canada contributes much to NASA's space program, and the Canadarms and the other devices that we have built are only surpassed by our personnel, who risk their lives each and every time one of them heads into space.

That was what the *Endeavour* was now doing. Public address announcer Navias had a particularly cogent observation to make just after the space-craft passed the seventy-third second of flight — a specific and telling reference to where the *Challenger* had been when it disintegrated in these same Florida skies. In referring to Barbara Morgan's presence on the shuttle, announcer Navias pointed out that she was now "racing towards space on the wings of a legacy." And she was indeed. Later on, in the heavens far above Idaho, a state where she once taught, Morgan conducted a lesson with students at a science centre in Boise, the capital city. Ronald Reagan would have been proud.

Endeavour took less than nine minutes to reach orbit, and by then it was 140 miles above Earth. This time, the announcer made mention of the entry to orbit, and proclaimed that "for Barbara Morgan and her crewmates, class is in session."

We all lingered for a while, watching the skies until there was nothing to watch. On the way back to the media building, I noticed Lisa Stark completing her last standup of the day for ABC News in New York. In a discussion I'd had with her earlier in the afternoon, the versatile and engaging reporter mentioned the torrid Florida heat that made her job particularly difficult that day, and then added, "But this launch was so important; I wanted to be here." All of her on-site reports were done outside, in excruciatingly humid conditions, and with a so-called "real feel" temperature of 108° Fahrenheit. "I hope all goes well with them," she said of the shuttle astronauts, "but you can never be sure...."[6] I know everyone there that evening felt her hesitation, even as we truly hoped for a successful mission.

About an hour after the *Endeavour* was safely on orbit, NASA officials held a press briefing at Kennedy. The packed conference room was a happy place, and all the questions were easy ones. It was left to Michael Griffin to summarize the work of the day: "This is one of the cleanest launches we have seen," he enthused. It was also one of the most widely watched in recent years.

After the shuttle had flown around the world a couple of times, I picked up my car at the foreign press parking area and left for my hotel.

The Commander's seat and flight deck interior of a space shuttle. Extensive electronic upgrades have been made to all of the shuttles. The first one that flew was *Columbia*, STS-1, on April 12, 1981.

Even then, most of the area highways were still jammed. Stop-and-go progress was an object lesson in how much interest there was, and is, in the space program and, in particular, the launch spectacle it provides. All of these vehicles carried individuals and families who had gone to see *Endeavour* go. And whether they had just witnessed their first or their tenth launch, they surely were not disappointed.

But what is the history behind this phenomenon, and how did Canada become involved in it? In order to answer such questions, it is necessary to look back at where we were at another time.

In so many ways, the last half century has brought about more change than any similar period in the history of humanity. Canada is, and has been, part of that change: in our people, our country, and our dreams for the ages that lie before us. We are, and will be, an integral part of a new space age that will be as exciting and full of promise as the one we have already known. But first, we need to understand and appreciate our contribution so far, and recognize those who have brought us to where we are today. The journey so far has been remarkable; the one that lies in the future will be even more so.

2

Crumpled Metal, Wires, and Waste

Canadian interest in space began long before Dave Williams went aloft in the spaceship *Endeavour*. In fact, since the beginning of recorded history, human beings have looked at the awesome beauty of the night sky and wondered about worlds beyond the stars. Our early ancestors revered the heavens, feared the heavens, and used myth and memory to deal with what they saw. In a sense, what we call astronomy today may well have been our earliest science. It just took us a while to give it a suitable name.

We know that the study of the heavens was important in Babylonia, ancient China, and India. Two thousand years before Christ, wise men from those, and other, areas of the world looked at the sun and moon and predicted eclipses. They also counted the stars, noted their brilliance, and gave them names. They mapped the skies, recorded what they saw, and systematically, painstakingly, and somewhat accurately, set down a legacy of learning for those of us who have followed. Early Greek astronomers determined that the Earth was round, centuries before the astronauts of today looked down from their capsules and took pictures of the Blue

Marble, as the Earth was dubbed in the first pictures of it taken from space.

Farmers, fishers, and sailors of old looked at the skies, and from what they observed, knew when to plant, fish, or head for the shelter of shore. Native peoples of every race had among their number respected elders who studied the clouds, made rudimentary predictions of what they portended, and, by word of mouth, passed the wisdom of their age to those who followed. Today, we do the same thing, but now we record, with ever-increasing accuracy, what we have seen and learned.

Because astronomy is "the study of the universe and its contents beyond the bounds of earth's atmosphere,"[1] the heavens were first studied in Canada long before it became a nation. In the early 1600s, explorers in the far north saw comets in the night sky, and eclipses that darkened the moon. Explorer Samuel de Champlain, and others of his time, knew wind and weather, but they also knew the stars that guided their ships in the face of that wind. Jesuit missionaries from France came here to spread the Christian gospel, and also made note of what they saw in the sky — including a lunar eclipse on a specific date: October 27, 1632.

We know that the first astronomical observatory existed, for a time, at Louisbourg, the French settlement on the coast of Cape Breton Island, even though all traces of the structure are long gone. Still, astronomy, its appeal and its value, was recognized elsewhere, and groups of like-minded individuals sought to advance studies in the field. This was particularly true later on, in Quebec City, Montreal, Kingston, and Toronto. An early observatory was built in Fredericton, at the University of New Brunswick, but the first recognized department of astronomy came into being at the University of Toronto in 1904. From that time on, interest in the science led to the establishment of permanent facilities elsewhere across the country.

An early reason for the particular emphasis on, and use for, astronomy in Canada came about because the north magnetic pole is located here. That, and the proclamation of two International Polar Years, resulted in increased interest in the north. During the second of these, beginning in 1932, scientists noted that radio waves "were bouncing off a layer of charged particles high in the Earth's atmosphere, and these waves were affected by Earth's magnetic field."[2] This observation led to further examination of the skies with telescopes and weather balloons. Since the beginning of the Second World War communication in Canada has become increasingly important. As anyone who has driven in mountainous terrain knows, radio reception in a car is often problematic. Signal strength rises and falls depending where you are: high on a promontory, low in a valley. A similar type of situation occurs in Arctic communities. For that reason, pioneers in Canadian radio realized that if they were going to give remote northern settlements a means to communicate on par with what was available in the south, workable techniques had to be found.

The quest led scientists to the realization that if they could somehow bounce radio waves, not off the ionosphere as they had been doing, but off something beyond it, the result could be improved communication anywhere on Earth. By the mid 1950s the idea of a man-made satellite was widely discussed, and its potential worth debated at length. Of course, at the time the problem lay in the launch. Just how would it be possible to shoot some kind of a payload beyond the atmosphere and into an orbit around the Earth? To accomplish that end, rocketry became far more important than ever before.

Experimentation with rockets of various types and sizes had been going on for years, in Canada and elsewhere. In this country, the bulk of the work took place near Churchill, Manitoba, and involved both American and Canadian personnel. The Bristol Aerospace Company in Winnipeg was heavily involved and operated in conjunction with an Ottawa-based organization called the Defence Research Board. However, rocketry had its origins much earlier and elsewhere, and almost always with connections to the military.

Indian and Chinese armies shot rudimentary rockets at their enemies a thousand years ago. In 1232, Chinese troops drove back invading Mongols with something called "solid-fuel fire- arrows."[3] In the late 1700s, Indian armies were using rockets that would fly for up to two miles. However, perhaps the best known rocket use is familiar to us, but in another context entirely.

During the War of 1812, an American military establishment called Fort McHenry, Maryland, was fired on by a British naval fleet. The attack, on the night of September 13–14, 1814, was staunchly defended and after a night of shelling the American flag still flew. A young Washington lawyer named Francis Scott Key witnessed the firefight, and even wrote a song about it. One hundred and two years later, Woodrow Wilson, the then President of the United States, designated Key's "The Star-Spangled Banner" as the official anthem of the United States. One of the lines of the composition refers to "the rocket's red glare." We have all heard it.

There have been several individuals in a number of countries who were important to the early development of rocketry, and the advances that led to artificial satellites and the space programs that we are familiar with today. Some of the more notable names from the past are: Konstantin E. Tsiolkovsky, Hermann Oberth, Robert Goddard, Wernher Von Braun, Sergei Korolev, and John Herbert Chapman.

Tsiolkovsky was the Russian who became his nation's "Father of Cosmonautics." He was responsible for scores of achievements, but will be remembered for his 1883 theory that rockets would work in the vacuum of space. As we know, he was right. Tsiolkovsky was also a bit of a dreamer, and some of the ideas he advanced came long before it would have been possible to bring them to fruition. For example, his "cylindrical spin-stabilized space habitat with artificial gravity and a space greenhouse with a closed ecological system"[4] was an idea he described in 1903! That was the same year the Wright Brothers flew their flimsy aircraft from the sand dunes at Kitty Hawk. Up until then, of course, no plane of any kind had flown, let alone a spaceship.

Hermann Oberth and Wernher Von Braun were both German, but the latter ultimately became an American after he was brought to the United States at the end of World War II. He was a highly skilled, inventive, determined man who had a major role in the design of the Saturn rockets that ultimately led to the successful American landings on the moon. But even with that great historical achievement, he would also be remembered as the man whose V-2 rockets wrought death and destruction on London, Paris, and elsewhere during the Second World War. On May 2, 1945, he surrendered to the Americans, who quickly put him to work on rocket development at their White Sands Proving Grounds in New Mexico. His new country chose to overlook his checkered past.

Robert Goddard was an American, Sergei Korolev a Russian, and John Chapman a Canadian. As is often the case, far too few of his countrymen

Courtesy Canadian Space Agency

The John H. Chapman Space Centre, headquarters of the Canadian Space Agency near Montreal.

have ever heard of Doctor Chapman, yet his contribution to space flight and space flight technology is unparalleled in this country. Today, the magnificent Canadian Space Agency site at Longueuil, Quebec, is named after him. He was a remarkable man.

Born in London, Ontario, on August, 28, 1921, the soft-spoken, quiet young man served overseas during the Second World War. At the time, radar was cutting edge technology. As a military officer, Chapman worked with radar at a time when its use was becoming increasingly important. The basic tenants of the science had been around, in theory, since 1864, but before Britain declared war on Germany in 1939, there were only five radar stations in the United Kingdom. Since radar could detect approaching enemy aircraft its use increased dramatically during the war. The device was particularly valuable during the Battle of Britain, when waves of German planes swept in from the English Channel, their bomb bays full.

Canada contributed much to the Allied effort in the field, and even built a Royal Canadian Air Force radar station at Clinton, Ontario. There, hundreds of men and women were trained in the use of radar prior to being shipped overseas to put their skills into practice. Clinton was chosen as the school site in part because the terrain around it was roughly comparable to coastal areas of Southeast England. The cliffs that border the Channel in Britain are much like the cliffs that overlook Lake Huron, near the Ontario towns of Goderich and Bayfield. Because John Chapman was from nearby London, he was quite familiar with such locales.

When the war ended, the young veteran returned home, and elected to stay in the field he knew so well. He obtained a degree in physics from the University of Western Ontario; then went to McGill for postgraduate work. In Montreal, his doctoral thesis was on "ionospheric radar echoes." In fact, "much of his research in the 1950s dealt with ionospheric physics."[5]

With his background, Chapman was ideally suited to assist when it became vital for Canada to pursue satellite research. In due course he went to Ottawa, where he became involved in ever-increasing leadership roles at what was called the Defence Research Telecommunications Establishment. He was working there when a landmark event in humanity's reach for space took place.

On October 4, 1957, the Soviet Union placed a satellite in orbit, and from that date forward, "the cause of human history was forever changed."[6] Now that rocketry existed to launch them, satellites sent aloft could, and would in time, become vital for telecommunications, weather forecasting, forestry, farming, ice mapping, and in countless other ways. They could also be used as weapons of war.

The possibility of satellites for weaponry was what had the most immediate impact in many world capitals — but none more so than in Washington, D.C. After all, the launch was at the height of the Cold War, and the satellite in the skies was Russian. The thing was called Sputnik, a shiny metal ball, less than two feet in diameter and weighing 184 pounds. It circled the Earth every ninety-six minutes, doing so 1,440 times before falling out of orbit and disintegrating. It made a "bleep, bleep" sound and had a flashing light: two things that surely irritated not only scientists and military planners, but most citizens of the United States who always assumed, and had been assured by their politicians, that their country was far ahead of Russia in technology.

Because the fear mongers of the time claimed that if they wished, the Russians could now put a nuclear bomb on a Sputnik and drop the thing on any part of the United States, ordinary Americans became deeply concerned. In fact, as one newspaper remarked more recently, Sputnik was "likened to Pearl Harbor and the 9/11 terrorist attacks — events provoking a national response to new dangers."[7]

Americans were also humiliated, and said so. Senator Henry Jackson, a crusty, blunt-spoken hard-liner from Washington State called the launch of the satellite "a devastating blow to the prestige of the United States as the leader of the scientific and technical world."[8] In fact, on the very day that Sputnik was launched, a prestigious scientific conference was winding up at the National Academy of Sciences in Washington. At the gathering, which included a handful of Canadians, a Vanguard rocket was on display. The thing garnered a lot of attention, and the "wholesome, eager-looking, and stridently self-confident" Americans who were displaying it assured conference delegates that the Vanguard "would be the first object made on earth to reach space." That was "beyond question."[9] No wonder they were chagrined — and shocked — when, later that same day, the Russians launched.

Yet, putting something like Sputnik into space was not exactly unexpected. For that matter, a hundred years earlier a visionary Frenchman imagined much more than just the launch of a satellite. A young lawyer, turned writer, named Jules Verne published a widely-read science fiction novel in 1865, in which he envisioned a rocket that would carry humans far beyond the atmosphere of Earth and, ultimately, to the moon. The book, one of many by the same author, was even called *From the Earth to the Moon*.

The writer, who is often referred to as the father of modern science fiction, was amazingly prescient. The rocket he wrote about was called Columbiad; it launched from Florida, went to the moon, and came down in the Pacific on its return to Earth. The work engendered as much admiration for its presumed precision as it did for its storyline. The former was not particularly unusual though, as the author spent long periods of time calculating in what direction, and how far and fast, his spaceship would have to travel. To ensure accuracy, he "even went to astronomers to have his calculations checked."[10]

Perhaps it is not surprising then that when the Americans went to the moon in July 1969, "Neil Armstrong named his Apollo 11 command module Columbia, after Verne's Columbiad."[11] He, and the two men with him, Michael Collins and Buzz Aldrin, had been given lots of advice about what to call their spacecraft, and they rejected hundreds of suggestions. However, because all three had read the Verne novel, their decision on the name was not really that unexpected.

But while the world was impressed, and the Americans shocked, by the Sputnik launch, another equally memorable one soon followed. This time, it was not just a satellite that bleeped and flashed as it circled the globe — a living creature was on board. On November 3, one month after Sputnik 1 was in orbit, a second rocket lifted off the pad at Baikonur. The living creature strapped inside was a doomed dog named Laika.

Before being placed aboard the spaceship, the little brown and white pooch had been a stray on the streets of Moscow. By all accounts, she was a docile animal, and did not seem to mind being strapped in the capsule with various censors hooked to her body. Photos were taken of her and through them millions of dog lovers, and others, developed an intense interest in the journey she would make. In due course, they would be stunned by her demise.

The officials in charge of Laika's trip assured the world that she was being properly cared for. They stressed that she had plenty of food and water for several days, and that she would not suffer from hypoxia, or lack of oxygen. Her vital signs were monitored at all times, and those readings were transmitted back to Earth. In all of the bulletins issued concerning the dog, none of them hinted that there might be a problem. Laika was said to be adapting well to space travel.

However, that was not the case. "Laika's cabin was cramped and overheated [and] there was no way to bring her down to Earth and no humane system had been installed for putting her out of her misery. Dehydration and heat stroke killed her."[12] Nevertheless, while Laika lived in space for less than a day, her journey was deemed a success. To a degree it was. At least rocket scientists now knew that an animal could endure a launch, and perhaps even an entire mission. However, the "entire mission" part still had to be proven.

Over the next few months, other dogs went aloft from Baikonur, but they too gave their lives for the cause. Finally, on August 19, 1960, the Soviets saw success. Strelka and Belka reacted well to space flight, and these two successors to Laika lived for years after their safe return to Earth. One of them, Strelka, was later bred, and in a curious footnote to history, a puppy she bore was presented to the children of U.S. President John F. Kennedy. That dog later had puppies of its own at the White House.

But whatever the Russian success launching animals, in 1957 the United States was still having trouble getting any kind of satellite into orbit. Their launch rockets failed again and again, with the ultimate embarrassment coming on December 6 of that year.

A Vanguard rocket, similar to the model displayed with such promise at the Academy of Sciences in Washington that fall, was moved onto one of the new launch pads at Cape Canaveral. The satellite, part of the United States Navy's developmental program, was intended to show the Russians up. The Vanguard was the way of the future; America would no longer trail in the space race.

Reporters from the major papers in the United States and Canada were there to cover the launch, as were news people from Europe and elsewhere. Television cameras were put in place, and the throngs of observers waited with ill-concealed anticipation for the big moment. Finally, all the essential preparations were done; the rocket was ready and all systems were go. The big moment would be televised across the nation.

Then the countdown began — and ended.

With a mighty roar and lots of smoke and flame, the rocket slowly began to rise — about six inches. Then the fuel that powered the thing exploded, and the Vanguard collapsed awkwardly back on itself. In short order, it was a crumpled mess of metal, wires, and waste that profoundly shocked those who were there and, through television, vast numbers of stunned viewers in homes and offices elsewhere. The moment, proclaimed one of the most well-known observers of his country's efforts to launch a satellite, had become "the image of the American space program."[13]

And so ended 1957; a year that had held so much promise, but in the final analysis undoubtedly benefited the Soviets more than the Americans. However, as we know, that would change. The United States would recover and go on to remarkable achievements in space. The next decade would see such advances, and Canada would play important roles in them. Within that time we would become only the third nation in the world to have our own satellite in orbit.

3
Trajectory of Achievement

There were several space-related developments in Canada during the early years of the Russian-American rivalry. Within hours of the launch of Sputnik, scientists at the Defence Research Telecommunications Establishment in Ottawa were among the first to detect and monitor the sounds of the satellite. Being able to do so was important, and it led to ever-increasing attention on the race to the heavens that was heating up between the two superpowers. Because of the great divide in world politics at the time, Canadian sentiment lay with the United States, of course, but what the Soviets had achieved was certainly recognized and applauded. Scientists such as John Chapman and his team understood the significance of Sputnik, but they guarded against the hysterical hand-wringing happening south of the border. In essence, the Canadian reaction was measured, objective, and understated. It did, however, lead to the view that perhaps a satellite for Canada might be a possibility reasonably soon.

Before that could happen the Americans had to move past their failures to launch, and get some kind of device into orbit. To their credit, they did not let the catastrophic explosion of the Vanguard lessen their resolve. As mentioned earlier, the Vanguard was the pride of the U.S. Navy, but now, to the chagrin of that branch of the American military, the first satellite that did make it into orbit was developed by the Army. It was called Explorer 1.

The thing was small, weighing only about thirty pounds, but the hopes of a nation were pinned on it. Chris Kraft, who later became the Flight Director for many of the future launches, vividly recalled that first one. It was January 31, 1958, only four months after Sputnik, when scientists working "in a dingy room [at Cape Canaveral] … flipped a toggle switch. Not far from that firing room, a Jupiter-C rocket spit flame and soared into the night sky. A few minutes later, Explorer 1 was a new satellite around the earth."[1] America was finally in space!

The relief felt by that nation, and indeed by the "Free World" was palpable. Scientists in Washington, London, Paris, Ottawa, and elsewhere realized that now the race for space had two participants, and either was capable of exploring beyond the boundaries of the Earth; perhaps even reaching the moon and beyond, in time.

Even though the United States was in second place at that point, it was the first to use its satellite for more than a propaganda piece. Explorer 1, rudimentary as it was, carried a small scientific payload. In addition to a thermometer of sorts that would keep track of heat variation between night and day during the orbits, and a microphone to record collisions with micrometeorites, a third instrument on-board was a kind of Geiger counter that would note and measure charged particle activity. This part of the package was what produced the most dramatic results. Data retrieved from Explorer 1 allowed scientists to determine that there were two layers, or belts, of "trapped radiation circling the earth. Measuring them provided the first detailed pictures of the shape of the planet's magnetic field; they defined the barrier that protects it from the deadly barrage of celestial radiation."[2] Subsequently, these belts were named after Doctor James Van Allen, the American astrophysicist who discovered them.

There were five important developments having to do with space-related endeavours that year. Four took place in the United States; the fifth in Canada. The first American one had been Explorer 1. The second, and much to the delight of the U.S. Navy, was the successful deployment of one of their own Vanguard satellites. Problems that plagued that machine had been gradually eradicated, and the first one was finally launched successfully on March 17. America now had two man-made "moons" circling the Earth. The propaganda value for the West was inestimable.

The third major component of space and exploration in America in 1958 was organizational, while the fourth involved an announcement like no other.

Around that time, and earlier, there had been ongoing discussions in the United States concerning the idea that some kind of bureaucracy might be necessary to keep track of space-related advances, and to lend some cohesion to the progress that was being made in the field. Particularly now, with the two satellites in orbit and several imaginative plans for the future, the impetus for a recognized structure became more widespread. Members of Congress, the Armed Services, the scientific community, and the President recognized the need. To that end, a new civilian organization came into being during the time that Dwight D. Eisenhower was in the White House.

This fledgling umbrella establishment was called the National Aeronautics and Space Administration, or NASA. As has been said by some, it started big and then became a colossus. Today it is a large and important bureaucracy, with links to every state in the U.S. and, in various forms, to places far beyond American borders. Thousands of people work under the umbrella, both directly and for companies across the world that supply myriad components for NASA and its related needs. Canada is among them.

The organization began on October 1, 1958, and a week later its senior officials made a move that was unparalleled at the time; both newsworthy and dramatic. On October 7, NASA formally approved Project Mercury, "to send a man into orbit, investigate his capabilities and reactions in space and return him safely to earth."[3] What was particularly unexpected about the announcement was that less than a year earlier the nation had still been struggling to get its first tiny satellite aloft. Achieving NASA's stated goal was going to take time, resolve, lots of money, and luck. American achievements were now being examined elsewhere; Ottawa was one of the places.

As soon as the Americans had succeeded in the launch of their two little satellites, John Chapman and his fellow researchers at the Defence Research Telecommunications Establishment began exploring the possibility of some kind of satellite for Canada. To some, even the idea was far-fetched. To others, it was felt that such a plan — a dream really — was not out of the question at all. Canada was a friend of the United States. Therefore, if Canadians could make a satellite, perhaps they could talk the Americans into launching the thing. In essence, that was what came about. The dream, one of the major steps in space exploration, was the fifth development of note in 1958.

At the time, however, both the Americans and the Russians had their sights on getting humans into space, not just satellites. More test animals would soon be riding rockets.

The first animals in space had been the dogs at Baikonur, and the Americans knew about those successes. How many dogs perished in Soviet space projects has never been revealed. Instead of using dogs as guinea pigs, the way the Russians had, Americans elected to experiment with chimpanzees. They were thought to be more versatile, and could be taught to do more things. Initially, forty of the beasts were slated to work with NASA. In due course the numbers were cut back, and then cut back again. Finally, six were deemed eligible for space training. Ultimately, one of the six, a male called Number 61, was selected to fly first. In deference to his soon-to-be flight status and probable fame, he was re-named. Henceforth, he would be known as Ham. Just prior to his arrival at the Cape, the dark-haired, shaggy-looking little primate had been kept at the Holloman Aerospace Medical Center, at an Air Force base in New Mexico. His rather prosaic new name reflected the initials of the medical facility.

By all accounts, the chimp was a quick study. He did well in training, was reasonably pliable, and quickly learned what he should and should not do once his launch day arrived. He was expected to open and close certain capsule switches on cue, as required by Cape Canaveral scientists. If he did as directed, he received a banana pellet. If he failed to flip a switch as expected, he was subjected to a painful electrical jolt in the soles of his feet. He soon understood how to avoid being zapped and he became quite adroit at opening and closing the necessary switches quickly and correctly.

When his day of glory came, Ham performed well, much to the relief of those who sent him aloft. He absorbed vastly increased G-forces, both on takeoff and re-entry. As one writer has said, somewhat irreverently, of Ham's journey in space: "his heart rate shot up as he strained against the force, but he didn't panic for a moment." As he surely figured out, "as long as he took it and didn't struggle, they wouldn't zap all those goddamned blue bolts into the soles of his feet."[4]

But there was a problem, and it was not Ham's doing. The capsule he rode in overshot its ocean recovery target by 132 miles. For that reason, it took the helicopter crews quite some time to find the actual splashdown location. Because of that delay, seawater poured into his capsule and Ham could easily have drowned. He survived, but he was no longer the docile little chimp that had been chosen to go aloft. While his rescuers worked as quickly as they could to free him he thrashed around and snarled, biting at their gloves, his harness, and anything else he could reach. Eventually, he was extricated from his water-heavy confinement and flown to the nearest recovery ship. Ham was now famous, and his name is always mentioned in the context of early American space experimentation and success.

And the chimp's achievement? Well, he proved to the scientists who sent him aloft that the time was fast approaching when a human being could ride a Redstone rocket. A Redstone gave Ham the flight of his life, and the feeling at Cape Canaveral was that such a rocket could take a man to space. To that end another Redstone was successfully tested and, finally, the one that would carry a man was brought to the Cape. It was trucked to and erected on Pad 5, at the south end of the Space Center property, a location that is deemed historic. There, almost on the Atlantic beach-front, and little over a mile from today's Port Canaveral cruise ship docks, preparations were made for the first launch of a human being into space.

As we now know, a Mercury capsule carrying a man was launched in Florida on May 5, 1961, but he was not the first to go to space. That success had gone to the Soviets.

On the twelfth day of the previous month, a twenty-seven-year-old Russian test pilot named Yuri Gagarin left the Russian steppes at

Baikonur, spent a total of 108 minutes in flight, and completed one circuit of the Earth. He achieved what American Alan B. Shepard was denied. Shepard made it into space twenty-three days later, and it was a Redstone that took him there.

Today, visitors at the Kennedy Space Center look at the Redstones on display and are invariably in awe of what Shepard actually did. They look at the rocket, compare it to launch vehicles now, and can hardly believe how small the thing is; only sixty-nine feet in length, a full sixteen feet shorter than a hockey ice surface is wide. It did, however, do what it was designed to do, and it at least partly restored American prestige at a time when it was sorely needed.

Al Shepard spent only fifteen minutes in flight, travelling 320 miles downrange from the Cape, but he proved to the designers that the Redstone was at least functional in the short-term. A similar rocket was used to launch a second American a couple of months later, but this time there was a problem — though it had nothing to do with the launch vehicle itself. Astronaut Gus Grissom's Redstone blasted off from Cape Canaveral on July 21, did a suborbital hop, and splashed down in what, up to that point, was a near-perfect mission.

That perfection was almost immediately forgotten when Liberty Bell 7, the tiny Mercury capsule he rode, blew a hatch and caused his near-drowning. Grissom got out into the ocean immediately, but no sooner had he done so than his spacesuit started to fill with water. He signalled desperately to alert the overflying chopper crews, but because they thought he was merely waving at them they waved back and then went to retrieve the Mercury before it sank.

The grappling hook from the nearest helicopter was affixed to the capsule, but because it was already partially filled with water, the hovering chopper lacked the power to do the hoist. Because of this situation, rather than risk a helicopter crash, the capsule had to be cut loose. It quickly sank to the bottom of the Atlantic, in some 16,000 feet of water.[5]

Meanwhile, Gus Grissom was floundering. He waved, shouted, fought the prop wash of the roaring helicopters, and cursed the flight crews who seemed determined to ignore him. Finally, the recovery teams realized that the astronaut was in danger. They hauled him out of the ocean and flew him immediately to the deck of the aircraft carrier *Randolph*.[6]

Grissom survived this incident but, sadly, would later perish with two others in a flash fire while training for a mission to the moon. He is enshrined today in the Astronaut Hall of Fame, just outside the Cape Canaveral space complex. Because of his historic early flight, and untimely death, Grissom is better remembered than most of the original seven astronauts that NASA selected for space flights.

While all this was happening, and indeed, even before anyone — Russian or American — had flown in space, John Chapman and his fellow scientists in Ottawa had assiduously pursued their dream of a satellite for Canada. In fact, NASA was barely operational when the Canadians submitted their pet proposal for consideration by senior members of the fledgling operation. Under the leadership of Chapman, the Ottawa-based Defence Research Board succeeded where no other nation had. The Americans agreed to launch a satellite for Canada, provided Canadians made the thing.

The agreement delighted Doctor Chapman, and he quickly assembled the key components of his team in order to build a machine, with the intention of studying the ionosphere. The proposal was formally approved by NASA on March 11, 1959. The understanding was that scientific information from the satellite would be shared, not only with scientists from the United States, but also with those from other nations who might benefit. At the time, the Americans most closely connected with getting a man into space already had direct knowledge of Canadian capability and resourcefulness. The communications antenna on the spacecraft in which Alan Shepard rode was of Canadian design, and had been built by de Havilland Aircraft in Downsview Ontario, now part of Toronto. The same antenna, called STEM, an acronym for Storable Tubular Extendible Member was also installed on the spacecraft used by John Glenn when he became the first American to orbit the Earth.

The antenna was a remarkable invention, perhaps best described by space technology expert Chris Gainor, as a device that could be "rolled up like a carpenter's measuring tape during launch, and when it was unspooled in space, it would form a tube and maintain its strength as it expended."[7]

As history has indicated, the Glenn mission was extremely important to America's advance in space, and Canada, with the launch of what was called Alouette 1 half a year later, also set out on a trajectory of achievement that continues to this day. From the 220 pound Alouette 1 that took flight from a Thor–Agena rocket launched in California on September 28, 1962, to the recent design and build of the amazing Dextre space station component, this country has maintained its expertise in a wide variety of aerospace endeavours. That is why Canadians are now allied with NASA and, to a lesser extent, with the Russians, the European Space Agency, Japan, and others in working and making advances in the forever frontier, the limitless boundaries of worlds yet unknown.

But along with all the technical achievements, innovations, and inventions, the human factor in any consideration of space is, without question, the most fascinating element of all. Even today, few men and women have gone beyond the Earth's bounds. And of those, only a tiny handful are Canadians. That is why their stories are unique. The following chapters will detail not only who they are and what they did and do, but will outline the developments that led to their becoming members of one of the most unique professions imaginable. They are astronauts — Canada's astronauts.

4
Canada's Membership Ticket

M arc Garneau flew first. As a Payload Specialist on the spaceship *Challenger*. He and six others were launched from Cape Canaveral on October 5, 1984. They spent eight days in orbit, flew almost three-and-a-half million miles, and landed on Runway 33 at the Kennedy Space Center at the end of their journey. The mission was a success, and Canadians everywhere rejoiced because one of our own was an astronaut. Others would follow, of course, but the young, handsome, perfectly bilingual, charismatic adventurer made us proud. From St. John's to Victoria, to Alert in the Arctic, the name Garneau would be recognized, celebrated, and forever enshrined as Canada's first in space.

But there were lots of steps, and yes, lots of heartache prior to Garneau, between Alouette and the decision by NASA to invite Canadians to be part of the astronaut program two decades later. At the time that Alouette was launched, both American and Russian accomplishments were decidedly primitive. Aside from Gagarin, Glenn, and a handful of others, few had flown. Back then, there were no women astronauts, no spacewalkers or space station, and landing on the moon seemed to be an impossible dream. Yet that dream had already been articulated, not long before the Glenn mission took place. On May 25, 1961, American President John F. Kennedy stood before a joint session of Congress in Washington, and said: "I believe this nation should commit itself to achieving the goal, before the decade is out, of landing a man on the moon and returning him safely to earth."[1] Kennedy's words energized his fellow citizens, and the message became a clarion call that demanded dedication, initiative, and the kind of creativity few nations ever show.

The tiny Redstones might have been adequate for the fledgling flights of Shepard and Grissom, but they were never enough for much more. Great rockets would have to be built, and they were. Moon landers would have to be invented, and they were. Capsule docking would have to be performed, and it was. And a million other things had to be understood, designed, tested, and tried before anyone was going anywhere near the moon, let alone to its surface and back. Somehow what was needed was imagined, invented, built, and used within the time that the young President mandated.

Canada and Canadians had a role in the race to the moon, although our contribution came, in part, because of a damaging event at home. On March 25, 1958, a new airplane called the Arrow lifted off the runway at Malton, Ontario, where Toronto Pearson International Airport is today. The plane, the most technically advanced fighter-interceptor in the world, was Canadian: designed here, built here, perfected here. It was also responsible for creating many jobs, particularly in Toronto and within commuting distance of the city. In a nation where reserve and moderation are so ingrained, the Arrow was a thing of pride.

Canadians had followed its development, knew of its potential, and on the day that it was rolled out for the public they came by the thousands to see it. Years later, when I interviewed Jan Zurakowski, the first person to fly the plane, he told me that the Arrow was "a magnificent aircraft ... far ahead of its time, and it showed that this country was in the forefront in aircraft technology — world-wide."[2]

And yet, for reasons that have never been completely clarified, Prime Minister John Diefenbaker ordered the Arrow development program stopped, and even directed that the half dozen planes already built be completely destroyed. His edict, issued on Friday, February 20, 1959, meant the immediate loss of over 14,000 jobs, the disruption of hundreds of families, and a loss to the country that was incalculable.

But the Prime Minister's decision was a big help to the Americans. Barely a month after the end of the Arrow, several senior-level NASA officials flew to Toronto and quickly scooped up all the best aeronautical engineers they could find. Foremost among them was the Arrow's brilliant Chief of Technical Design — Kamloops, British Columbia native, James Chamberlain. He, along with Owen Maynard from Sarnia, Ontario, and twenty-three others who had built Canada's showcase aircraft, needed jobs and were enticed by the American offer. Reluctantly, they realized that they had to leave Canada to find work in their field, but were relieved that the opportunity in the south came when it did. As Christopher Kraft, the eminent Flight Director at Mission Control, noted in his memoirs: "in one bunch, we got engineers who could make major contributions to getting us into space."[3] Kraft, who would later become the director of the Johnson Space Center in Houston, was always lavish in his praise for those who came from Canada when they were sorely needed. He knew them all, befriended most, and saw to it that they had a chance to use their skills for the race to the moon.

Then, one after the other, there were new types of spaceships, and new types of rockets to launch them. The largest of these, the mighty Saturn V, was integral to NASA's Apollo program, and is still the star of the show in one of the visitor galleries at the Kennedy Space Center. Another Saturn has been erected at a rather unlikely spot, but it does draw attention to its place of origin. At a Welcome Center, a half mile from the Tennessee-Alabama state line, on southbound Interstate 65, the 363-foot-tall Saturn is impossible to miss. Thousands of Canadian snowbirds see it as they drive towards the Gulf each winter. Nearby Huntsville, Alabama is where the rocket was built.

It was that machine that carried the men on their journey towards the moon. On July 20, 1969, when Neil Armstrong and Buzz Aldrin touched down on the Sea of Tranquility, their lunar module landing gear was Canadian, built by Heroux Aerospace of Longueuil, Quebec. Canadians assisted in astronaut training, spaceship design modification, tracking at Mission Control, moon rock laboratory testing, astronaut suit improvements, and countless other areas of involvement.

When Armstrong, Aldrin, and Command and Service Module Pilot Michael Collins returned from the moon, they had to be quarantined for three weeks in case they brought lunar germs back with them. And while no such germs ever materialized, a medical doctor was quarantined as well, to ensure that the fliers were germ-free. The doctor, Edmonton-born flight-surgeon Bill Carpentier, was highly respected by the three spacemen, and by other astronauts he dealt with later on.

On later moon missions, the astronauts involved did not just collect moon rocks; they attempted to bring back geologically significant ones. In order to be able to identify specific types of these, they trained in a variety of places. The Cinder Lake region near Flagstaff, Arizona was explored. So was a location in Texas, and another at Ries Kessel, in Bavaria. But, of particular interest to Canadians, time spent in Sudbury, Ontario was important. "In July, 1971, astronauts John Young and Charlie Duke got their first look at Sudbury's ancient impact breccias in situ. Eight months later, they would sample lunar rocks of this origin in the vicinity of their Apollo 16 landing site."[4] Astronauts also examined geological formations at the Canadian military testing grounds just outside of Suffield, Alberta.

On January 5, 1972, U.S. President Richard Nixon told the world that NASA was about to embark on the building of what we know today as the space shuttle. Already the euphoria over the moon landings was beginning to subside, and the Apollo program that put men on the lunar surface was winding down. The last astronauts landed on the Taurus-Littrow region of the moon in late December of that year, and spent seventy-five hours exploring, and collecting soil and rock samples there. Then they flew home, and no one has been back since.

On November 9 of the same year, with a publicity flourish, NASA launched Anik 1 for Canada. With it, we became the first country in the world to have our own geostationary communications satellite. Other versions of Anik followed, and their operations brought "network radio, TV and improved telephone services to Canadians living in the north."[5] At long last, no matter where they lived, people in this vast land would be able to communicate more readily with each other. Just as, in a different arena entirely, we were already communicating well with NASA.

Because of Canadian participation in the various programs that led to the Americans reaching the moon, our presence and expertise were known and appreciated by the most senior officials at NASA. That was why, when it became apparent that some kind of lifting and placement machine was going to be needed on the space shuttle, NASA turned to Canada and asked us to build the thing. And even though the Americans still insist on calling it a Remote Manipulator System, to Canadians the robotic arm has always been the Canadarm.[6] We invested heavily in the device, and perfected the engineering that went into it. The first one was launched on the space shuttle *Columbia* on November 13, 1981. From the start, the system exceeded expectations for its usefulness and reliability.

"Canadarm was our membership ticket into the space program," Chris Hadfield told me, "and the invitation to fly came because of it. It represents a huge Canadian involvement."[7] The invitation that Colonel Hadfield referred to arrived in Ottawa on September 29, 1982, and directly led to Marc Garneau's presence on *Challenger*, just two years later. His flight came at the end of a whirlwind preparation period.

Unlike many of the men and women who would later become his colleagues, Garneau did not give much thought to becoming an astronaut prior to his being selected as one. In part, this was because in the early years Russians and Americans were the only ones who went into space. Young Canadians who might have been interested, even if they thought themselves qualified, simply disregarded the idea because they were not from a nation that was involved. However, as soon as NASA extended its invitation to Canadians, the possibility of riding the shuttle struck a chord across the country. In fact, over four thousand people claimed to want the job that Garneau got. Ultimately, six would be short listed from a group of the best nineteen candidates.

Both men and women applied; people of all ages, from a variety of backgrounds, with dreams that ranged from the practical, to the positive, to the purely whimsical. Some had hope, many had no hope, and a few had no obvious understanding of what was involved, but even the intentions of those applicants had to be assessed. Narrowing the field took time, soul-searching, and political sensitivity. From the outset it was made known that at least one person out of the final six had to be bilingual in both French and English. It was also hoped that one or more would be female, although the selection committee flatly rejected the implication made by some in the media that tokenism came into play in the final decisions.

The process took half a year and in the end the group selected was arguably the finest potential astronaut corps from any nation, anywhere. They were: Roberta Bondar, Marc Garneau, Steve MacLean, Ken Money, Bob Thirsk, and Bjarni Tryggvason. For the purposes of this book, only the stories of those lucky enough to have flown will be expanded upon. That is why there are no chapters on Ken Money, Robert Stewart, or Michael McKay. Money was ultimately passed over when Bondar was named to a shuttle crew. Stewart and McKay were picked to be astronauts during a second selection round in 1992, but both ultimately cited personal reasons for declining to continue in the program.

At the time he applied, Marc Garneau was a naval officer with a Ph.D. in electrical engineering. His father, Andre, was a retired army general from a French-Canadian background. Because Marc's mother Jean was English-Canadian, it came as no surprise that their thirty-five-year-old son spoke both languages with an equal facility. While he had all the positive qualities necessary, Doctor Garneau admits that he also was a lucky man to be where he was when the advertisements for astronauts first appeared. In earlier years he had used some of his vacation time in July, but during that month in 1983 he was working in Ottawa when he happened to see the notice in a local publication. "It was just fate," he told a reporter later, "I saw a career ad for Canadian astronauts in the paper. If I had taken my holiday in July, I would never have applied."[8] But he did apply, along with all the others. Then they waited. And waited.

Finally, on Saturday December 3, six of the nineteen short-listed candidates were told they were astronauts, Garneau among them. Then, barely three months later, he was picked as the first to fly. By that time his life, and the lives of his new colleagues, had changed forever.

5
A Flash, a Roar, and a First

The training program for astronauts is tough. It is extensive, intensive, time-consuming, and relentless in its expectations. In the months, and often years, that lead up to a launch, astronauts endure as much physical, psychological, and academic probing as anyone on Earth. They also fly planes, scuba dive, skydive, make speeches, and deal with an often cynical media and an indifferent public. They also have to stay in shape, absorb reams of data pertinent to their new calling, adjust to unfamiliar and often contradictory work demands, juggle impossible schedules, and try to figure out just where they fit in with their colleagues, instructors, mentors, and bosses. On a personal level, they have to maintain family stability, attempt to give as much of themselves as humanly possible at home, and work out compromises within their particular circumstance. There is little time off, and few chances for relaxation or reflection.

All of Canada's new astronauts were subject to such pressures, and in order to survive had to find within themselves organizational skills they may not have known they had. To begin with, they were all expected to repair to Ottawa when training began, but as soon as they gained some understanding of their new roles there, other demands ensued. To complicate matters, "with the first group of astronauts, the training plan evolved on the fly."[1] For Marc Garneau and Bob Thirsk, his backup, it also meant a move to Houston in the weeks before Garneau actually flew. Prior to that, the two, and their colleagues, had undergone training at McGill University in Montreal, and at the Defence and Civil Institute of Environmental Medicine (DCIEM) in Toronto. The group also flew to the Cape to tour the Space Center so that they all would have at least some familiarity with the place before being named to a shuttle crew. In most cases, such selections were far in the future, but at least they had seen KSC. They did not, however, see a launch. While the Canadians were in Florida the shuttle mission they had hoped to watch was scrubbed. It eventually did go, but two months later than planned.

Then there were the constant demands for public appearances at civic events, university commencements, elementary and secondary school assemblies, building dedications, charity functions, and even shopping mall extravaganzas. All of these meant media interviews, autograph signing, posing for pictures, handshaking, and sitting through boring and lengthy introductions at service clubs and sports banquets. These functions were deemed necessary because the space program was expensive and the audiences at such gatherings were members of the taxpaying public.

While Garneau's training was unpleasant at times, he had to endure it and go on. So did the others. In Montreal, "the astronauts practiced and refined the procedures for the various SASSE (Space Adaptation Syndrome Experiment) tests — vestibular — ocular reflex, sensory function in limbs, awareness of the position of external objects and proprioceptive illusions."[2] In Toronto, motion sickness and astronaut susceptibility to it was a prime factor in the training provided. While it is known that many astronauts feel queasy in space initially, preventive measures before flight help to counteract the phenomenon.

At Defence and Civil Institute of Environmental Medicine (DCIEM), two machines, both of which were devised by Ken Money prior to his being selected as an astronaut, had to be ridden by each new candidate.

Courtesy Canadian Space Agency

Marc Garneau is a veteran of three space flights: STS-41G in 1984, TS-77 in 1976, and STS-97 in 2000.

One training device was little more than a chair attached to the end of a pole. The person in training had to sit in the chair and move his or her head up and down while the chair on the pole was spun faster and faster. A similar situation would be a high-speed, sharp turn on a roller coaster. The main difference, however, would be that the roller coaster turn is over in a flash. The ride at DCIEM lasted several minutes.

The second machine at the Toronto lab flipped the astronaut-in-training heels over head, again and again; similar to that first, over-the-top ride on a Ferris wheel. In the lab, however, the over-the-top part was backwards, and it lasted until the person taking the test began to feel ill. All of the new astronauts sensed the onset of motion sickness as they were tossed over and over, but none threw up.

All six trainees also had to be subject to High Altitude Indoctrination (HAI) to prepare them in the event of an inflight emergency resulting in an oxygen supply failure. Anyone flying a fighter jet undergoes the same measure — including the author prior to taking-off in one of Canada's F–18 aircraft. The new astronauts were placed in an altitude chamber where they experienced the equivalent of being in a plane. At high altitudes the atmospheric pressure decreases. Should there be a pressurization failure, the effect on the human body can be injurious, leading even to death. Without oxygen, we die.

While at a simulated high altitude, having been deprived of oxygen the astronauts were handed a clipboard and told to draw various shapes: a circle, a square, a triangle, and so on. As they did so, they were told to note how they felt without oxygen, in case the same thing ever happened to them on the shuttle.

While the exercise can be stressful, it is both enlightening and, to a degree, rather amusing. In my own case, the first few shapes I drew were recognizable. By the ninth or tenth though, I felt that I could not force my hand to do what my brain directed. When I utterly failed to draw a simple square, I grabbed my mask and breathed the sweetest-tasting pure oxygen imaginable. The astronauts did the same exercises, and learned from doing them.

During the time that the recruits were undergoing the various tests, they were helping to put into place a series of experiments that Garneau would conduct while he was on the shuttle. His role on the flight was as Payload Specialist, and his job was to work through several scientific exercises that would be beneficial for study during and after the mission. A couple of these were taste and smell tests, in which he tried to decide if a substance seemed sweeter, or perhaps less spicy, in space than it did on Earth. Another experiment dealt with spatial awareness: Garneau attempted to tabulate the ways in which human movements in space differed from the same actions on Earth. He was also tasked with taking photographs of the aurora borealis and a supposed reddish glow around the shuttle that had been noted on other missions. He also conducted tests to measure pollutants in the upper atmosphere, a field that has become ever more critical in the ensuing years.

Practice for such activities continued when he and Thirsk got to the Johnson Space Center (JSC) near Houston. The Space Center is a huge, sprawling enterprise a few miles southeast of downtown Houston. Driving to it today is horrendous, as a visitor has to cope with drivers whose only speed is fast; multi-level, inadequately-marked freeway interchanges; routine tailgating; and lane changes where signalling is apparently frowned upon. No wonder large, rather alarming signs indicating accident reporting centers seem to be everywhere. There is little doubt that such places are needed. This was the city where Garneau and Thirsk headed, albeit in a somewhat quieter era.

The Space Center itself resembles a university campus, with identical sand-coloured buildings on all sides. The architecture is dated now, but when the place was built in the early sixties it reflected the style of the time. And while there is little beauty in any of the structures, they are functional, and what happens within them is known about around the world. Building 30, housing today's Mission Control, is large, windowless, and utilitarian. Within it, the original Apollo Era Mission Control Center is quiet now, as befits the historical site that it is. However, it was the

The Apollo era Mission Control Center in Houston. This was the control room used for all of the early space flight operations, including the moon landings and the flights of the first Canadian astronauts. It is now a designated historic site.

Courtesy John Melady

nerve centre for Garneau's mission and it operated until mid-1996. This is where controllers worked on all the moon missions, and where nerve-racking decisions were made to bring the *Apollo 13* crew home after their spaceship exploded in flight.

Once they arrived at JSC, Garneau and Thirsk began a sped-up version of the training they had been doing for months. Now the schedule was a non-stop whirlwind of requirements — everything from meeting the personnel they would fly with, to getting measured for spacesuits. They also were able to experience weightlessness for the first time, without going into space.

NASA uses what they call a zero gravity training aircraft to give astronauts a taste of weightlessness, allowing them to more easily adapt to space once they get there. The plane used at the time, a Boeing KC-135, had most of its seats removed, leaving a large padded area where astronaut-trainees could work. The plane climbs, then pitches over and dives rapidly; during the pitch-over, everyone on board becomes weightless for about half a minute. The climb-dive process is repeated again and again, and each time the short period of zero gravity occurs. When this is happening, the astronauts are not buckled into seats, allowing them to float around, and learn how to adapt to being weightless in space. Because the wild ride on the training aircraft sometimes makes the trainees sick whatever plane is being used is derisively called the Vomit Comet.

The *Challenger* crew flew to Florida on October 2, 1984. Finally, launch day was at hand. The gleaming white spacecraft was already in place on Pad 35A. In the three days before it would hurtle into space, the vehicle and every system on it was checked again and again. Then, during those last critical hours before launch, the tanking procedure was completed: the necessary fuel was in place.

The astronauts themselves were paraded before the media for a last, pre-flight photo op, during which Marc Garneau spoke briefly to reporters in both French and English. His comments were succinct and expected, admitting that he was looking forward to the mission, which he called "a big adventure." He also mentioned that he was "the fortunate Canadian who gets the chance to go up in the shuttle."[3] The next day, his remarks appeared in most newspapers in Canada, reflecting his status as a national hero.

Garneau's English-born wife, Jacqueline, was in Florida to watch the launch, as were Simone and Yves, their eight-year-old twins. Garneau's parents and his brothers, Philippe and Braun, were also there. Their actions and words were duly reported by many of the more than one hundred Canadian media representatives who were accredited by NASA for this milestone mission. One person who had been invited, but did not attend, was the nation's new Prime Minister, Brian Mulroney. He had been in office for less than three weeks. His absence at KSC prompted speculation that he and his government were uninterested in the space initiative, and were particularly critical of its cost. Some even speculated that the Garneau mission would be the last for Canada. As we now know, the conjecture was unfounded, but was reflective of the constant funding concerns that space programs encounter — from all the countries involved.

The day before launch, Jacqueline, Marc, and his parents attended a beach barbecue together, but only after the visitors agreed to medical checks to ensure that they would not pass so much as a slight head cold to the quarantined crew. The children were not invited in case they were in any way ill. NASA had a long-standing precedent for this. Fourteen years earlier, astronaut Ken Mattingly was bumped from the *Apollo 13* moon mission because, although not sick himself, had been exposed to German measles by the children of a crew member. He eventually flew, but had to be replaced on his intended mission. Being pulled from it apparently caused "the worst depression of his life,"[4] and NASA did not want a repeat of such a problem with another astronaut.

For several days before launch, the *Challenger* crew had been going to bed early and getting up early. This was a not-very-successful attempt to ready them for flight day. When the time finally came they were called at 3:00 a.m., had breakfast, and were on the way to Pad 39A an hour later.

The spaceship lifted off at 7:03.

For miles around, the flash and roar of the shuttle seemed to be the only thing alive in the darkness of that Florida morning. Those who watched and listened were awed by the spectacle, made even more dramatic by low-level clouds that reflected a white then glorious pink afterglow, as *Challenger* hurtled through them. Inside the spaceship, the rumble of the ride and the lurch into space were both anticipated and intimidating, and the crew went along for the ride. In less than nine minutes they were in space, and NASA officials were already calling the launch "the slickest in memory."

On this flight, as with all others, one of the first realities astronauts dealt with was the absence of gravity. Things float around, people float around, and perceptions of "up" and "down" are meaningless. There is also the constant effort to avoid motion sickness. The problem can be annoying, lasting, and at times troublesome, particularly during the execution of inflight duties — regardless of the astronaut's job. A sick Commander is just as uncomfortable as a Payload Specialist. Although Garneau admitted that he felt nauseous at the outset, he did not get sick at any time during the flight. He was more interested in the magnificent scenery outside. "That first view is spectacular," he said, "that's something you've been dreaming about seeing for the longest time. The thing that struck me was how crystal clear the colours were."[5]

Later, in interviews after the successful mission, Garneau mentioned that at the time of the liftoff he knew his heart was beating faster than usual, and admitted to being apprehensive with all the noise and vibration. The feeling passed quickly though, and the nervousness was supplanted by seeing the beauty of the Earth and the unforgettable scenery outside the shuttle windows.

Overall mission purposes, apart from Garneau's own duties, included an Earth Radiation Budget Satellite deployment and spacewalks by two of the American astronauts; one man and one woman. As well, a Canadian-based IMAX camera was used on the flight. Anyone who has seen the large format films produced using such equipment comes away impressed. Invariably, moviegoers marvel at how real the images are — so real that viewers feel like they're there.

The first hours of the 132 orbits were important to the crew as they accustomed themselves to being in space. The shuttle interior is not large, and after being in a mock-up of the crew module in Houston, I was amazed that several people could work, eat, and sleep in such close quarters. There were seven astronauts on board and their living space was undoubtedly cramped. The mid-deck, or largest area of the crew module, is only eight feet wide, five feet nine inches long, and six feet nine inches high. Much work was done there, and everyone had to adapt to the close quarters. To facilitate this the seats used for takeoff and landing, and other non-essential inflight equipment is stowed out of the way once the shuttle enters orbit.

On this mission, Robert Crippen was Commander and Jon McBride was the Pilot. Unlike the airlines of the world, no one on a shuttle is called a co-pilot. There were three Mission Specialists, two of whom were women, and another Payload Specialist with Garneau. The crew worked well together, and the Canadian on board was impressed by their professionalism. This was particularly true about Crippen, flying his fourth mission in space. At a press briefing in Houston after the flight, Garneau called Crippen, "One of the most fantastic people I've ever met."[6] High praise indeed from a normally reserved and rather taciturn colleague.

That reserve was the only thing that journalists would fault Garneau for during the mission. The man was just too quiet, they groused, forgetting that his first responsibility was his work on the shuttle, not placating reporters. From time-to-time Garneau did give the press people material for their dispatches, even if the first reported comment was all business. In a message to Mission Control, the focus of his performance on the spaceship comprised his entire message: "I'd just like to let Canex personnel know I've completed all the objectives today," he said. "As far as Viset is concerned, we completed alpha and bravo, charlie with the first and third with the first and third translations plus sections delta, echo, gulf with camera delta and bravo. Only camera A did not provide visible video and also section A alpha was repeated after the rules of the ERBS satellite."

The reporters on the ground were left shaking their heads as to what all this meant, but they duly recorded those first words by a Canadian from space.

Later there were other messages that were more personal. Marc and Jacqueline celebrated their eleventh wedding anniversary during the time he was aloft. The press found out about this, and many worked a reference to the occasion into the stories they filed. Garneau, in turn, wished his wife a happy anniversary, but ever-picky, some members of the media noted that he was two days late in doing so. For her part, Jacqueline sent her husband a few words that she admitted were of a romantic nature. However, whatever she said was too much for the public relations people at NASA. They deleted most of it.

Shortly after the takeoff from the Cape, the Canadian Broadcasting Corporation informed radio listeners that Brian Mulroney would be making a phone call to Garneau, to congratulate him on his historical flight. For whatever reason, the usually talkative parliamentarian elected not to make the call, and the CBC had to issue an apology. "The report was released without the confirmation from the Prime Minister's Office," a CBC spokesperson told the audience. Instead, a phone call to the *Challenger* from the then President of the United States was made.

In a four-minute communication with the shuttle, Ronald Reagan went out of his way to praise Garneau, and in so doing alluded to the Canadarm that had become such a vital part of the spacecraft machinery. "I'd like to say hello to Canada's fine astronaut," said Reagan. "With all there is to do on this mission, I know Cripp (shuttle Commander Robert Crippen) appreciates having three strong Canadian arms."[7] Garneau thanked the President and added, "It's a great honour for me to be aboard. I'm having an incredible time and it's just great to be here."

Garneau was pleased and somewhat surprised one afternoon when his American crewmates presented him with a gift on the Canadian Thanksgiving Day. The "gift" was actually a replacement meal. Instead of the standard spaceflight dinner, Garneau was handed a tinfoil packet of turkey and gravy. He laughed and thanked his fellow fliers, but declined to comment on the quality of the culinary delicacy.

As each day passed, crewmembers settled into a routine that was both workable and efficient. They were awakened each morning by musical selections from Mission Control, and went about their duties with professionalism and dedication. All felt that there were just not enough hours to accomplish everything. On the morning of the second full day in space, Houston greeted the crew in French with the words, "Bonjour mes amis," and followed this with the playing of "What a Feeling," from the movie *Flashdance*. There is no record of how that particular selection was received: it was played at 4:10 a.m., Houston time.

While other crewmembers went about their own mission assignments, Garneau applied his full concentration to his work. In general, the experiments he was involved with were successful. Among the photographs he took were some highly detailed images of Montreal. The scientists who received them felt that they were the best ever taken of the city from space. When Canadians elsewhere learned of this, they felt pride that one of their own was contributing in this new and exciting venture.

The people of Kingston, Ontario were among them. As the shuttle flew over their city one night, Kingstonians switched their lights as a greeting to *Challenger*. The gesture could be seen from space, and later Marc Garneau called it "touching." His appreciation for his country as a whole was evoked in his replies to questions from reporters on the ground, at the mid-point of the mission. Because the shuttle was travelling at over 17,000 miles per hour, and circling the globe every ninety minutes, each time over his homeland was brief. "We are over Canada only twenty minutes, [and the] view is absolutely extraordinary," he said at one point. "My country is fantastic," he added in French. "I believe we are very lucky to be Canadian and to have such a big and wonderful country."[8]

That feeling was reciprocated. In an editorial just prior to the mission, the *Toronto Star* mentioned the Canadian pride in the history-making endeavour:

> With his scheduled launch tomorrow, Canadian astronaut Marc Garneau will advance a Canadian space program that has been soaring aloft for two decades. Ours is a less exotic program than that of the Americans and Soviets. But it's a solid one. Garneau may be flying on American wings, but Canadian hearts and best wishes are flying with him.[9]

Then, all too soon as far as the crew was concerned, the mission was over. There were the usual concerns about weather in Florida, and the possibility of a scrubbed landing there.[10] However, the skies cleared at the Cape, and the shuttle returned without incident. Shortly afterwards, NASA, where precision is all-important, announced that the total time for the mission had been eight days, five hours, twenty-three minutes, and thirty-three seconds. But whatever the total, the fliers were happy to be back safely and their families were even happier to see them. A couple of hours after touchdown in Florida, the astronauts flew on to Houston where their loved ones were waiting — in the rain.

The much-needed wet weather did little to dampen spirits in Texas, and the flight of the thirty-five-year-old Garneau was deemed to have been a complete success. Sadly though, not every mission would be as successful, nor end as happily. In fact, before the next Canadian flew, a terrible accident threatened to bring an end to the entire space program.

6

A Moment of Hope and Pride

The cold in Florida was unusual, unexpected, and brutal. Foot-long icicles hung under the upturned shuttle. The steel catwalks closest to it were sheets of ice, and the metal tubing of the access tower was covered with grime. A bitter wind blew in from the sea.

That morning, January 28, 1986, feverish preparations were being made to launch the spaceship *Challenger* and the seven souls on board. The cold was of concern, and the ice team, in their final inspection twenty minutes before the scheduled liftoff, recommended a delay so that the coating could melt. The hold was for two hours, but against the better judgment of some, at 11:38 a.m. Eastern Time "the engines howled and the stack rose. The ambient temperature was 15 degrees colder than [for] any previous shuttle lift-off."[1] The cold was still worrisome, but time constraints called for a launch on this day, and urgency trumped safety. This bird would fly.

But not for long.

The *Challenger* was destroyed in a matter of seconds. Even before the smoke and steam had dissipated at the launch pad, pieces of the shuttle were falling into the ocean. Overhead, great billowing exhaust clouds were white and terrible against the blue of the winter sky. On the ground, the thousands who watched were aghast, disbelieving, and stunned. Some screamed, some swore, and some prayed as the pyrotechnics of horror played out above them. The shuttle was gone now, and so were the seven men and women who flew in it. Left behind were shattered families, second guesses, and plenty of blame.

Four months later the members of the committee that President Reagan had appointed to investigate the disaster finally presented their findings. They had determined that the shuttle was lost because of technical design problems that were made more pronounced by the unusually cold temperatures at the time of the launch. The investigative body also criticized personnel within NASA, and two shuttle contractors. The American manned space program was stopped for thirty-two months.

In the immediate aftermath of the tragedy, an intensive search began along the nearby shoreline and in the adjacent waters of the Atlantic. The tiniest pieces of plastic, metal, and unidentifiable debris were collected with the utmost care. Even more care went into the recovery of the remains of the astronauts, all of which were located offshore, still within the crew module. In fact, *Challenger's* crew compartment stayed intact, trailing a mass of umbilical lines like streamers as it continued to climb to 65,000 feet. Ten seconds after the explosion, the crew compartment went into free-fall. It slammed into the Atlantic two minutes and forty-five seconds after the blast at 207 miles an hour.[2]

To the shocked observers that day, crew survival seemed to have been impossible; and it was. However, it took NASA more than seven hours to confirm the deaths, and the organization was widely faulted for such insensitivity. Then there were the funerals — all of which were heart-wrenching affairs. Even as the deceased were being laid to rest, the hunt for debris continued; thousands of sections of the shuttle, and as many parts of the rocket boosters were located. After months of examining and probing them for clues to the cause of the accident, the tangled remnants were entombed in two deserted intercontinental ballistic missile silos at the Cape Canaveral Air Force station.

When the work was completed in January 1987, massive concrete caps were lowered into place, sealing the wreckage in silent darkness.

Today, visitors to this lonely, isolated part of the Air Force station see only the weathered concrete caps. Rusting metal, loose gravel and tall grass give the area a desolate look as ocean breezes swirl across this gravesite.[3]

NASA suffered as an organization from what had happened. In fact, "the explosion shattered public perceptions about the seeming routine nature of spaceflight, and about NASA's infallibility. Within the agency, amid shock and bewilderment, came the sudden realization that they had all been living too close to the edge."[4] But, to give credit where it's due, in time, like a Phoenix rising, the shuttle program resumed. There were new safeguards, new personnel, new warnings, checklists, and failsafe procedures that filled volumes. NASA, its employees, contractors, and hundreds of brilliant people whose efforts had brought about the Space Transportation System (STS) or, to most, simply the shuttles, all rededicated themselves to their work. They used another space shuttle, positioned it for flight, and on September 29, 1988, watched it soar. Even though it had flown prior to *Challenger*, *Discovery* was the first launch following the accident. Five astronauts were on board, and the shuttle mission, designated as STS-26, was a resounding success. Even though someone has calculated that "shuttle astronauts are given a one percent chance of dying on a flight,"[5] there was relief beyond measure when the spaceship touched down at Edwards, four days after it left Kennedy. For the next several years, the shuttle operation continued without pause. Finally, during that time, on January 22, 1992, the second Canadian flew. This time it was a woman. Her name is Roberta Bondar.

Bondar was born and raised in Sault Ste. Marie, and almost from the time she could talk had a keen interest in the natural world around her, and in the skies above the pine trees of her northern Ontario home. A natural athlete, gifted student, and the possessor of an unshakable dream of going into space, Bondar did everything in her power to achieve the end she sought. She worked hard, got top marks in every course she took,

Roberta Bondar is a veteran of one space flight: STS-42 in 1992.

Courtesy Canadian Space Agency

and built a resume more impressive than most people could ever imagine. After elementary school, secondary school, and eighteen years of university, she finally convinced those whose decisions counted that she was not just ready to fly; she was as accomplished as any astronaut around.

Some felt that she was passed over at first because she was a woman, and while that may have been the case, she insisted, and rightly so, that she was never chosen for her mission because she was a woman. She had all the necessary credentials and more, and proved her detractors wrong at every turn. By the time she was ready to fly, Bondar had a Bachelor of Science in zoology and agriculture, a Master's degree in experimental pathology, a Ph.D. in neurobiology, and a pilot's license. She was also a medical doctor who, at the time that she was short-listed to be an astronaut, was treating patients at a Hamilton clinic.

When Bondar flew, for the first time the shuttle carried a science facility, generally referred to as Skylab, in its huge payload bay. This International Microgravity Laboratory would ultimately fly several times, facilitating a myriad of scientific experiments done in the absence of gravity. Bondar's primary role on *Discovery* was to do science experiments and, like Garneau before her, she was a Payload Specialist for the trip. Also like Garneau, she had confidence in her ability to perform the tasks set out for her.

In her work on this flight she helped to advance the purposes for which the shuttles were designed, but perhaps not in the same way as had been anticipated. Years earlier, Robert Freitag, an advanced programs planner at NASA, thought of the shuttle as primarily a people mover with a great future. "I'm convinced that by 1990 people will be going on the shuttle routinely," he claimed with confidence, "as on an airplane."[6] As we know, of course, that prediction was off base, but without doubt, by carrying a science lab into space the shuttle was proving its worth in a different, but perhaps equally important way.

The lab that was carried on the Bondar flight was an inventive, complicated, and futuristic piece of machinery. Developed by the European Space Agency, the microgravity laboratory in the payload bay was connected to the shuttle mid-deck by a tunnel. Astronauts traversed the tunnel to enter the lab, and worked there in far more comfort than they had in the crew module. For one thing, because the space was bigger, they were much less crowded. Here, often in shorts or slacks, T-shirts, and socks, they performed experiments on behalf of scientists around the world. In Bondar's case, some two hundred scientists had a direct connection with what she was doing. She was in constant contact with Ken Money, her backup astronaut, who monitored her every action from the Marshall Space Flight Center in Huntsville, Alabama. Money probably had more understanding of her work and routine than anyone else on Earth. As backup, he had done the same training as she, had followed the same step-by-step, intricate requirements, and in essence, completed the exact same procedures. However, he did not go into space then, and to his great disappointment, did not get to do so in the future either. He left NASA in the months after the Bondar mission.

The amazing laboratory on the shuttle provided Bondar with the facilities and room to work through a total of forty-three experiments involving many of the biological specimens brought along in Spacelab. Included in the menagerie of travellers were "frog eggs, fruit flies, lentil roots, slime moulds, spores, stick insects, bacteria, human cells, mouse cells, nematode worms, yeast and plants."[7] Unfortunately, the fruit flies perished on the voyage — killed by sterilizing chemicals on the containers in which they were packed. Then there were experiments on humans; in this case, fellow astronauts. Tests were done on the inner ear, the eyes, the heart, the back, the lungs, and the legs. Exercises were performed with the subject wearing headgear and sitting in something called a Microgravity Vestibular Investigations chair, or on a sled that moved on rails. In order to maintain their fitness level in space, the crew all worked out on a modified rowing machine in an effort to prevent muscle deterioration.

And the astronauts "grew" on their flight. Because there is no gravity pulling down on the body, the spine stretches in space, and astronauts find that they are temporarily taller than they are on Earth. This situation often induces back pain, and while the problem is well-known now, it was still under study on the Bondar mission. Bondar too grew taller in space. In an answer to a question following a recent address, she said that she was about an inch and a half taller after her flight. "For the first time in my life," she joked, "I was taller than my sister." The audience at the gathering laughed when she added: "I even had a photo taken of the two of us to prove it."[8]

The work ethic of astronauts is well-known, and for her part, Roberta Bondar worked flat out, in an attempt to complete every experiment possible within the time constraints of the mission, eating, sleeping, and exercising. In a reference to Marc Garneau and his work habits, one journalist wrote that an astronaut's schedule "is usually a workaholic's dream."[9] The same could just as readily have been applied to Bondar. Despite getting sick on the shuttle and having "to pull out the vomit bag,"[10] those in a position to know of Bondar's work habits praised her for her dedication and achievement.

So did American President George H.W. Bush and Prime Minister Brian Mulroney. The latter, despite deciding not to talk to Marc Garneau when he was in space, did so with Bondar. For his part, President Bush said: "I just wanted to say how pleased we are with you representing Canada. I think it's a wonderful thing and I think it's a wonderful way it shows the strength of ties between our two great countries." Mulroney congratulated Bondar "on becoming Canada's second space traveller," and added that the flight had "captured the imagination" of her fellow Canadians. And it surely did.

Roberta Bondar's mission was covered as extensively by the media as the flight of Garneau. Anticipating this, she reciprocated in kind by letting the press know what she was doing and seeing on the flight. Her every word was reported and several newspapers ran editorials praising her. Typical of those was one story published the day after STS-42 left the pad:

> What Canadian throat didn't have a lump in it as the shuttle *Discovery* blasted off cleanly yesterday, rockets spitting and crackling into the Florida sky? In the hardest of times, it was a moment of hope and national pride as the remarkable Dr. Roberta Bondar became the first Canadian woman astronaut in space. Beyond her many duties in the cramped International Microgravity Laboratory, Bondar has a larger assignment — to give science in Canada the human face it so badly needs.[11]

For her part, Bondar also had words to say about her voyage — particularly the beauty of the Earth that she saw from the windows of the shuttle. "No one can prepare you for the overwhelming experience of being in visual contact with the Earth from space," she wrote later. "It is an awesome adventure."[12] Then she added: "The view is incredible. The rich blues of the ocean, the exquisite turquoise of the Great Barrier Reef, and the soft white snows of winter are soothing to eyes that have been engrossed in demanding scientific duties."[13]

These scientific duties were what captured the imagination of many on Earth —particularly school children. As he talked to Bondar from Washington, President Bush was joined in the Oval Office by a group of teachers and their students, who represented an organization called the

Astronaut Roberta Bondar speaking at a function in Toronto in the spring of 2007.

Courtesy John Melady

Young Astronauts Program. The school children who were there asked Bondar, and four other shuttle crewmembers, questions about weightlessness, whether it was better living in space than on Earth, what the Earth looked like from space, and about the weight of things in the shuttle. One of the American astronauts did a demonstration on the video link, and used the Canadian IMAX camera on board to illustrate his point. He told the youngsters that the camera weighed between 110 and 120 pounds on Earth, but that the instrument would float freely in orbit, and that Roberta Bondar, who was no heavier than the camera, could operate it easily.

A somewhat similar teleconference took place at the Ontario Science Centre in Toronto. There, in an auditorium packed by about a thousand elementary school students and teachers, Bondar's face flickered on the giant screen as she talked about the wonders of being in space. The audience was enthusiastic, and the educators in the crowd felt that the event was a great way to interest youngsters in science. Several students in the building that day claimed that after seeing the astronauts floating around inside the spaceship they now wanted to become astronauts themselves.

Similar thoughts were expressed across Canada — not just in an auditorium in a big city. In Sault Ste. Marie, Bondar's hometown, excitement over the mission ran high. Everyone talked about it, about Bondar, and about the fact that this event would "put the Sault on the map." A fifty-foot-long banner dedicated to the town's space hero was hung at the local arena. A plot of city land would be called the Roberta Bondar Discovery Playground, and a couple of new office buildings were named after her. A life-sized snow sculpture of *Discovery* was created, and a local concession stand sold Bondar Burgers, Discovery Dogs, and Shuttle Sausages. In essence, the mission was a complete success and everyone said good things about it — well, almost everyone.

At the end of the first full day in space, an unfortunate news item in the *Toronto Star* drew the ire of many, not only in the Sault, but across the country. The piece was written by a staff reporter for the paper, who was covering the mission from Huntsville, Alabama, because backup Astronaut Ken Money was there.

The story grew out of the reporter's interview with Doctor Alan Mortimer, Chief of Life Sciences for the Canadian Space Agency. Mortimer mentioned all of the things that astronauts had to do in space before work in the science lab could begin. He mentioned putting away the spacesuits, the stowing of seats in the crew module — which, while needed for takeoff and landing, are unnecessary in orbit — and the checking of various systems to ensure they are in proper working order. He summarized all those activities by saying that "housekeeping needs to be done,"[14] and that Canada's first female astronaut — and the only woman on the shuttle — would be doing such work. Then, to compound matters, the headline writer for the front page article made things worse by captioning it: "Bondar spends hour tidying up shuttle."[15] Published with the story was a photo of Bondar herself.

The paper was barely on the street before there was reaction. Angry letters, phone calls, telegrams, and even hand-delivered complaints flooded into the paper. Before *Star* staffers could deal with one irate reader, several others were waiting in line to vent. All were outraged at the sexist reporting. Some said they intended to cancel subscriptions, others demanded retraction, and many said they would never read the paper again. All were outraged.

"I am angry and disappointed that the editors at the Star would allow such a stereotypical, out-dated and glaringly chauvinistic headline," wrote April Goodis, from Oakville, Ontario, in a letter to the editor published on January 25, 1992. She added: "I have a 5-year-old daughter who looks to me as a role model, to provide guidance, encouragement and perspective towards pursuit of the goals she sets for her life. Shall I therefore, suggest she hone her housekeeping skills should she intend to pursue a career in the space program?"

The same day the paper published other letters, one among them from Jessica Burton, of Brampton, Ontario, who observed: "I don't believe I have ever seen a more stupid lead on a story. I assume that is your idea of cute." Cynthia Maxwell, from Willowdale, Ontario insisted that the story reflected "a missed opportunity to show that Canadian women can compete and succeed on a world scale." While Penny Lamy, of Toronto, was sure that "if a member of the opposite sex was hurtling through space on a mission of scientific discovery, that insulting headline would never have been printed — especially above his photograph. Acts such as this perpetuate the abhorrent myth of a woman's place and ability."

Three days later, on January 28, Jennifer Ramsay more or less summed up the reaction of many when she wrote, with what can best be termed cutting irony: "What progress we have made: from barefoot and pregnant and in the kitchen to light housekeeping duties in space!" Ramsay's message to the editors was unequivocally pointed: "Your decision to highlight Roberta Bondar's 'housework' over micro-gravity experiments and pharmaceutical research makes me sick."

Fortunately, the newspaper did respond — but without actually saying that such a thing would never happen again. Rod Goodman, the ombudsman for the *Star*, devoted an entire column to the matter. He mentioned that there had been more than 150 complaints, and that at the paper itself "an air of near disbelief pervaded parts of the newsroom after papers came up from the presses. Some reporters and editors wrote stinging notes." Those in-house complaints, coupled with messages from the public "were sprinkled with such adjectives as disgusting, sexist, revolting [and] outrageous. Men and women were both represented in the hostile reaction."[16]

Though Goodman was obviously displeased with what his paper had done, in a curious deflection of blame he said that other papers made mistakes too. That week, he argued, "the Toronto *Sun* trashed [Bob Rae] the Premier of Ontario with the front-page epithet What a Boob!"[17] Whether Goodman's rationalizing satisfied his own readers is unknown, as are the numbers of those who gave up on the paper entirely. The *Star* continued to cover the shuttle story, of course, even as the mission was winding down.

High above the planet, *Discovery* and its crew were making preparations to land. Things used — or not used — on the flight were securely stowed prior to re-entry. One such item was Roberta Bondar's glasses. As an unexpected result of being in zero gravity, she realized that she had no need of the glasses she wore on Earth. However, early in the mission, when she took them off, the glasses had floated away. Fortunately, she did locate them prior to landing.

All in all, STS-42 was as successful as it was historic from the Canadian point of view. As Roberta Bondar has written of the experience: "I have witnessed wonders from space [and] I will always think of the earth as never before, cherishing the sense of awe that this flight has inspired within me."[18]

The shuttle landed on concrete Runway 22 at Edwards, at the close of a momentous mission that had seen it circle the Earth 129 times. Now, Canada's second space pioneer, our first female astronaut, was safely home.

7

No Longer a Trainee

Steve MacLean is a patient man. He was picked to be an astronaut during the first selection process in 1983. Back then, he was unsure of himself, inexperienced, somewhat naïve, and perhaps less aggressive than his colleagues. He remained on the sidelines when Garneau flew, was still there during the Bondar mission, but finally got his chance in late October 1992 — nine years after he was first called an astronaut. He accepted the wait, but deep down knew what fabled American space veteran Gus Grissom had meant when he said that "you're not an astronaut until you go up."[1] You could parade around in the fancy flight suit, but until you flew, you were just like the rest who waited; to Grissom, and to the old pros of his time, you were still a "trainee."

Yet every trainee has to start somewhere, and MacLean did. He got his initial flight, and then waited, and waited — for fourteen more years. Then he flew again. On his first mission, STS-52, he was a Payload Specialist, the lowest level of flier in the NASA pecking order. However, by the time his second flight was ready to go, he had become a highly trained

Mission Specialist who would conduct a risky spacewalking assignment that would be seen on television screens in Canada and across the world. But as inexperienced as he was on the first flight, he performed well on the mission — as well as anyone on board — and was lauded for doing so. The shuttle he rode then was *Columbia*, and he and the crew completed several operations and flew over four million miles, in all. When the spaceship returned to Kennedy, almost ten days later, MacLean was no longer a trainee. Gus Grissom would have been proud.

Steven Glenwood MacLean was born in Ottawa, on December 14, 1954, the son of a scientist at the National Research Council. At one point, after the loss of *Challenger* when all flights were grounded, father and son got the chance to work together in the nation's capital. In fact, Steve's father expressed initial reservations about his son leaving

After describing his second mission in space, Steve MacLean answers questions during a press scrum at York University in Toronto. Doctor MacLean is a former champion gymnast and a graduate of York. He is now president of the Canadian Space Agency.

Courtesy John Melady

the security of the scientific community for the relative unknown of the astronaut corps. However, the son's dreams were in space, and he intended to follow them there.

The future astronaut received his elementary and secondary school education in Ottawa, then went on to attend York University in Toronto, emerging with a Bachelor of Science in 1977, and a Doctorate in Astrophysics in 1983. During an interview for this book, he referred to York as "home," and is proud of his link to that school. He has also been the recipient of several honorary degrees from other universities, along with many medals and awards of various kinds from several sources.

Doctor MacLean is a gifted athlete, and even when he's just walking across a floor he exhibits the grace of a person who is at one with his physique. He is not a big man; he is slight, and at first glance seems almost fragile. He moves quickly, smiles easily, and speaks with enthusiasm about the work of other astronauts, but with self-effacement in reference to himself. He only refers to his athletic skills if specifically asked, but it was those skills that led to his notable achievements as a young man, before the idea of becoming an astronaut took hold with a firmness that could not be denied.

At one time, MacLean was one of this county's finest gymnasts. In fact, he was not only champion university gymnast in 1976, he was also on the Canadian National Gymnastics Team for three years in succession. These levels of excellence are remarkable, particularly because he was in grade eleven before he decided to give the sport much of a try! He took it seriously though, and for years he trained several hours a day. Hiking is another passion, and it became an important part of his life — so much so that when he was in his twenties, he hiked to the base of Mount Everest. At other times, MacLean has unwound by playing tennis, or relaxed by going canoeing.

Steve MacLean was twenty-nine and single when he made the cut as one of this country's first astronauts. At the time, he had his Ph.D. from York and was doing post-doctoral work in laser research at Stanford University in California, when a buddy called and said they were looking for astronauts in Canada. Those selected would be flying with NASA, and ultimately would work out of Houston. That is still the situation today, and though we have a well-respected space agency and facility at Longueuil, Quebec, the bulk of an astronaut's training is in Texas, and most live there.

Nevertheless, there is one aspect of training that has distinctive Canadian roots: language instruction in our two official languages. All Canadian astronauts are expected to be able to speak and write in both English and French, perhaps not with the same facility as Marc Garneau, but well enough to do media interviews, make speeches, and interact with the public in both tongues. While he is fluently bilingual today, MacLean had to work on his French in order to achieve the ends desired. So did Roberta Bondar. Her French lessons had a positive, if unexpected, connection to MacLean. He met his future wife while she was teaching Bondar French.

In the post-*Challenger* era in Houston, there was much soul searching by many astronauts — and others — about whether to stick with the space program or not. That was to be expected. For his part, as we all know, Steve MacLean remained interested, committed, and ready to go as soon as the shuttle flights did. This dedication paid-off for both him and Bjarni Tryggvason, who was MacLean's backup for STS-52. Tryggvason would fly later.

MacLean's training for his first mission accelerated markedly in the months immediately before launch. He and Tryggvason were in Houston then, and while neither was a Pilot or Mission Specialist, they went through long and rigorous days when they were pummeled, pushed, and prodded in more arduous ways than had been the standard before the *Challenger* disaster. There was water survival, fire suppression, and ascent profile simulation, in which the fledgling fliers were exposed to G-force acceleration, all in preparation for anything that could happen on launch day. They spent hours and hours in a shuttle simulator, a full-scale version of the shuttle flight deck. There, within an ugly, closed, awkward-looking contraption above gears, pipes, wires, hydraulics, and vent pipes, the motion of a shuttle and its capabilities were imitated. Inside the simulator were all the same dials, switches, displays and controls, a portable oxygen system, and so on that are on the shuttles. There were also simulator visual scenes replicating the surrounding terrain at Kennedy, White Sands, and

Edwards, as well as Moron and Zaragoza Air Bases in Spain, and Istres in France. The latter three are what NASA calls Transoceanic Abort Landing (TAL) sites that can be used in case the orbiter had to make an "unscheduled landing if one or more of its three main engines failed during ascent into orbit, or if a failure of a major orbiter system, such as the cooling or cabin pressurization systems, precluded satisfactory continuation of the mission."[2] Such an emergency would be declared between two and a half, and seven and a half minutes after a shuttle left the pad. While MacLean would have no part in making such a landing, he and the five others on board had to know how to get out quickly in the event of a problem. As Bjarni Tryvaggson explained, "they're throwing malfunctions at us, and the crew has to sort these out. These guys [and one women: Tamara Jernigan] do not go five minutes without something going wrong. Everything has to be responded to like its real."[3] Such preparations are all part of getting ready for every mission.

Long before his training time in Houston, MacLean had married, and was lucky enough to have his wife Nadine in Texas with him. Sharing their lives by that time were a son and a daughter, both of whom were still very young. The fact that the four could be together was a bonus, made possible because Nadine was on maternity leave from her job in Ottawa. By being close to Steve in Houston, she had a much better idea of the training regime he had, while he, in-turn, was able to help her with their children. As it is, astronauts have to be away a lot, so any opportunity just to be home is welcomed. For her part, Nadine was, and is, completely supportive of her husband's work. Over the years, the times away have been an accepted part of his job. But in the lead up to launch, the job is extremely time-consuming and time on the job is critical. This was particularly true because of the inclusion of a prototype for the Space Vision System (SVS) on this flight. Testing the SVS was MacLean's primary role on the mission, even though there were several other experiments that he had to do as well.

The Space Vision System is uniquely Canadian and it "links computers with television cameras shooting targets on satellites or other vehicles to provide real-time computer images of where the Canadarm is in relation to the object it needs to grasp."[4] The system had been under development for some time, and MacLean felt a profound obligation, not only to those who built the thing, but to those who were flying with him, and to his country to do the required tests precisely, completely, and satisfactorily. Being the kind of man he is, there was much self-imposed pressure involved.

Because the SVS would be employed on the flight deck, MacLean would work up there, and that location in itself was a tribute to his expertise. Because the technology was state of the art at the time, and looked as if it would become increasingly necessary on future missions, its inclusion and performance on this one was critical. With SVS, the Canadarm operator would be able to see what he or she was doing, even if the arm was maneuvering something that could not be seen from the flight deck itself.

But the testing of the SVS was not the only important development in Canadian space initiatives in 1992. The flights of Bondar and MacLean were the highlights of course, but that same year the Space Agency opened a campaign to hire new astronauts. After weeding through the resulting 5,330 applications, four new candidates were selected on June 8. Then, less than a month later, one of the newcomers, Chris Hadfield, along with Marc Garneau, became the first Canadians selected to be Mission Specialists. Their training began a month later in Houston. However, the focus in the latter part of the year was on the MacLean mission.

A few days before the launch, the shuttle crew completed their long period of rigorous training at the Johnson Space Center and flew up to Cape Canaveral. They went into quarantine soon after arrival, and from that time on immediate family members who had been checked by NASA physicians were the only ones who saw them. Nadine and her parents, along with MacLean's parents, and scores of his friends had come down from Ottawa and elsewhere to witness the launch. For most, it would be the first, and perhaps the only time they would see such a spectacle, and none wanted to miss it. Included in the group were several former gymnasts who partied the night away before the 11:16 a.m. scheduled blastoff time.

The group, who came to Florida by chartered bus, even sported identical blue T-shirts with MacLean's picture on the front, and a lighthearted, uniquely Canadian message on the back: DOCTOR STEVE MACLEAN, BOLDLY GOING WHERE NO GYMNAST HAS GONE BEFORE. I KNOW HIM, EH. One of the organizers of the group, Rob Robichaud from Toronto, summed up the mood of his mates, and said, referring to MacLean, "We're having a party. If he can't make it, fine."[5] MacLean did not show, of course, and nor did they expect him to, but the exuberance of his friends was certainly not diminished because the guest of honour was not there. They were there to see their gymnast colleague speed from the Earth as he blasted off on the shuttle *Columbia*; though it would happen two hours later than originally intended.

Severe cross winds at the emergency landing strip necessitated a delay in the departure of the spacecraft, but after NASA personnel decided that the risk was minimal the countdown was permitted to continue.

When the blastoff happened, Nadine watched it with the other wives from a vantage point atop the large, white, flat-topped launch control building. The dark glasses she wore hid her tears of apprehension and pride, as the man she married disappeared on the Roman candle hurtling before her. In her arms that afternoon was thirty-month-old Jean-Philippe MacLean, who looked at the rocket with a child's concern and asked: "Papa coming back? Papa coming back?" Nadine assured her son that his father was returning, knowing in her heart that, unfortunately, the youngster would likely never remember this day. Later on, she spoke to a reporter about the experience of watching the launch: "It was over before I knew it," she said. "It was just an amazing, amazing feeling. I don't think I'll ever feel anything like that again."[6]

That same afternoon, Steve MacLean's father, Paul, who had also watched the launch, explained his reaction to a reporter: "You get a tremendous feeling of being completely innocent and naïve when you're faced with such a magnificent demonstration of power that man has been able to create in sending off a rocket," he said. "When [it] exploded off into a tall column of burning flame — everyone just got up and cheered, as if it was a strikeout or another home run for the Blue Jays."[7]

The baseball reference was particularly appropriate that October. The Toronto Blue Jays were indeed getting both strikeouts and home runs at the time. They were in the thick of their first World Series run, and both

Steve MacLean and his father were avid ball fans. In fact, Steve took both a baseball and a Jays cap with him into space. Once there, he and the rest of the crew were given scoring updates as the Jays and the Atlanta Braves competed for the sport's top prize. When Toronto won it all on October 24, MacLean was jubilant, as was one other crew member, although that person's identity was not revealed because it meant that one of the Americans on board was hoping that the team from the United States would lose.

But even baseball and the Toronto World Series win were unable to compete with the adrenalin rush of reaching space. That first test of real weightlessness — not the vomit comet kind — prompted MacLean's remark that it was "phenomenal" to be where he was. "This is great!" he enthused.

The first few days of the mission were devoted to the various experiments that were to be done apart from the work on the SVS. That was left nearly to the end of the voyage, largely because it was feared that some of the Canadarm maneuvering might cause shuttle vibration that could have affected other delicate scientific work that was being carried out. When the Arm was in operation it was used to deploy an Italian satellite from the payload bay; a large, round, mirror-surfaced device that was designed to reflect laser beams from Earth. It was called LAGEOS, and would be used to study the tiniest shifting of the Earth's crust and, if possible, help to predict earthquakes. NASA saw to it that the launch received as much publicity as possible, ostensibly to help counter ongoing criticism that space missions were too costly. Things done, or deployed, on any mission helped to justify those costs. If whatever was in the payload bay could improve life on Earth, then the Space Agency wanted the world to know about it. The Italian satellite was an example of this.

Apart from his testing of the SVS, MacLean worked with a Canadian-designed device called a photo spectrometer in order to obtain "high quality information on the state of the ozone layer and on the various gases and dust particles that contribute to destroying and creating ozone."[8] Rays from the sun and the moon were analyzed, in order to determine the way light permeated the atmosphere. In addition to MacLean's work, there were ongoing American experiments involving other crewmembers. One of these, done by Tamara Jernigan, proved to be of particular interest. She rode a stationary bicycle for intense periods at a time, in an effort to figure out how to ensure that astronauts could stay in shape, particularly as space missions got longer.

During the *Columbia* flight there were the usual shuttle to Earth telecasts, and in general they went off without a hitch. When he was on camera, MacLean could not resist wearing his Blue Jays cap. Doing so led to good-humoured joshing and Steve's admission that he had won some money because of the Toronto win. He proved to be quite knowledgeable about baseball, and at one point gave a brief dissertation on player strength, pointing out that his favourite team had more pitching depth than Atlanta. In the end, and with the Jays win, his evaluation proved to be quite sound.

But it was the pressure MacLean put on himself that made him the ideal astronaut as far as NASA and the Canadian public were concerned. He did his best, didn't shirk responsibility, and performed in the way he told a reporter he would. In an interview a week before the mission, he noted that he would be "working up there. I'm not really being observed. All the experiments I have are hands on. It's not like taking photos and coming back and having them developed. I have to do something to make it work. As a professional, you can't ask to be in a better situation."[9]

And MacLean did work — often non-stop for hours at a stretch. Sometimes he even forgot to eat, and laboured through lunch periods. "We're working a sixteen hour day," he told the CBC during one of the televised interviews from space. "We don't stop until it's time to go to bed, and for two or three days, I hardly got anything to eat because we were so busy." He added that Canadians often had more experiments to do, per flight, than the Americans, but that this was because flights open to Canadians were so limited in number. Nevertheless, MacLean's efforts did not go unnoticed, and *Columbia* Commander James Wetherbee had one of the American Mission Specialists assist with some of the work. In doing so, the Commander stated that the Canadian was "probably the busiest" member of the crew. "He has the most experiments and he's performed just superbly." Then the former naval officer added: "It's obvious Canada has sent us her best."[10]

Another side of MacLean's personality that emerged partway through the mission involved an earthbound interviewer asking crewmembers if they had noticed, or believed that little green men might inhabit the universe. Amid laughter, the answers that were given were all over the map. Some said that we might not be alone; that there might be life elsewhere. No astronaut admitted seeing anything unusual though, and one or two said they doubted there were humans of any kind, anywhere but on the Earth. When MacLean was asked for his views, he jokingly replied that before he left home his mother told him that once he was in space, he should not talk to anyone he did not know. He did not say if he had encountered any strangers up to that point of the mission. In any event, Commander Wetherbee said he would keep an eye out for other life forms — just in case.

During the same interview crewmembers were asked about their taste in music. MacLean did not get a chance to reply before the end of the broadcast, but those who knew him recalled that one of the things he intended to take with him was a cassette produced at York. On it were selections by the university music faculty.

While Steve MacLean was busy trying to accomplish as much as possible on the shuttle, his wife Nadine was making the most of the time until he returned. She had finally forced herself to relax and worry less about the mission, largely because there was nothing she could do if problems did arise. When she had the chance to do so, she watched the telecasts from the shuttle on NASA TV, and realized one morning that she saw more of Steve during those few days when *Columbia* was in orbit than she sometimes saw of him when he was on Earth. Now and then, when the telecasts were on, little Jean-Philippe would walk over to the television, touch his father's image on the screen and say "look — Papa!" to his mother. At eight months, baby Catherine is still too young to know what was going on.

Close members of Nadine's family had been with her in Florida, and their presence was appreciated. Together they relaxed, shopped, visited Disney World, and swam at the beaches. Midway through the waiting time, she flew to Houston, to Mission Control, in order to more fully appreciate her husband's work activities, and to talk with scientists and others who were closely connected to work on the shuttle. She was also able to contact Steve through the Johnson communication facilities. Later on, she spoke of what she most wanted to hear from him when he returned. "I'm looking forward to hearing his impressions — the real ones. The private ones," she said. "Hopefully he'll take time to write down or record his impressions."[11] MacLean did have lots of memories of what he had seen, heard, experienced, and enjoyed. Then, all too soon it seemed, the mission was over. The SVS tests were highly successful, and the man most responsible for them could finally relax and return.

Nadine was back in Florida in time to embrace her husband shortly after the shuttle landed. After all of the flight debriefing, and mandatory NASA and Canadian Space Agency activities, Steve, Nadine, and the children were finally able to spend time together — on a holiday in Hawaii.

8

Our Top Gun in Space

On July 20, 1969, when Neil Armstrong stepped onto the moon and into history, a nine-year-old farm boy at Milton, Ontario, was watching the event on television. Despite the grainy, unclear black and white image, the youngster was enthralled by what he saw. He watched as a figure in a bulky white spacesuit stood on the footpad of the lunar module, hesitated momentarily, and then, as he lowered himself onto the lunar surface, spoke the words that were flashed around the world: "One small step …" The boy could hardly contain himself as the image flickered before him. A man was standing on the moon, outside, up in the sky, and far, far away. He was the first person in the history of the world to do such a thing, and it was so exciting!

"That's the moment I decided I wanted to be an astronaut when I grew up," Chris Hadfield recalled. "It seemed the most amazing thing imaginable."[1] For the millions and millions of people who were watching the event, the moon landing seemed as if it was a science fiction epic concocted by Hollywood. We saw those first few steps; the planted flag, the ghostly images of utter desolation, the stark blackness, and then the bouncing, bumbling "moonwalking" performed by first one, and then two men.

That scarcely-believing boy at Milton saw what we all saw, and from those moments on, he has spent a lifetime reacting to the events of that night. The image of the first man on the moon became an image in history, like Pearl Harbor, the assassination of Kennedy, or the falling towers of the World Trade Center. Those of us who remember some, or all, of those events will have them with us always, but while they are implanted in our consciousness, we take them no farther. Chris Hadfield did — from the moment he saw Neil Armstrong place his foot on the dust of an alien world. The landing on the moon inspired Hadfield to a point that almost everything he has done since had its roots in that instant of history. "It was one of those significant emotional events," Hadfield explains, "I just knew I could do it."[2] And he has.

A superior student, a top-notch air cadet, then a university graduate, a pilot, and ultimately, an astronaut, Hadfield has excelled at everything he has done. Some would say he is driven, and perhaps he is, in the sense that once he sets a goal for himself, he becomes focused on achieving it. His intention is not just to be good at what he sets out to do; he wants to do it to the best of his ability. And there is nothing wrong with that.

After breezing through elementary school, he did even better at the secondary level. When he graduated from Milton District High School in 1977, he was an Ontario Scholar, meaning that he was among the best of the best. The same pattern continued all through the rest of his years of formal education. He did not just graduate from the various universities he attended, he did so with honours.

As a young teenager he joined the air cadets, wore a uniform, learned to march, salute, and fly — and again excelled at all of these aspects of the organization. Joe Vick, a cadet warrant officer at the time, remembers Hadfield as "a fine young man. He was popular, well liked, and had a pleasant personality. He worked hard and had a real drive to succeed. He was also a natural leader, very serious, polite, and much smarter than anyone out there, including myself."[3]

And no wonder. Hadfield is a member of Mensa, the organization that "welcomes people from every walk of life whose IQ is in the top

The author with veteran Canadian astronaut Chris Hadfield, at Cape Canaveral.

2% of the population."[4] That disqualifies most of us, but certainly not Chris Hadfield.

After he graduated from high school, Hadfield decided to take a year off and see as much of the world as he could in that time. He and a buddy flew overseas, and once there took city transit, walked, hitchhiked, and rode trains across much of Europe, including some of the countries that at the time, and even today, were not premiere tourist destinations for Canadians. When he was included in the dozen or so personalities in the Maclean's magazine Honor Roll (*sic*) for 1994, the writer who compiled his profile mentioned that Hadfield "marked his 18th birthday on a train in Bulgaria," and by 1994 had lived in "Canada's three westernmost provinces and Quebec, in California, Maryland and Texas."[5] He has since resided in several other places, including Russia, and in addition to being able to converse with ease in either French or English, writes, reads, and speaks fluent Russian. "Although I still would never attempt to discuss philosophy in Russian," he told me.

After he explored as much of Europe as time would allow, Hadfield joined the Air Force, principally because he saw it as a stepping stone on the path to becoming an astronaut. Once in the military, his first degree was in Mechanical Engineering from Kingston's Royal Military College (RMC). Then there was postgraduate work at the University of Waterloo, and a Master of Science from the University of Tennessee in 1992. He has several honourary degrees as well.

During his final year at RMC, Hadfield married Helene Walter, his girlfriend from high school. Over time, the couple became the parents of two boys and a girl — all of whom, like their parents, are achievers. When asked about them today, their proud father loves to tell you what they are about. One lived in England and India, and went to school in both countries; another is currently working on a scholarship at a university in Osaka, Japan; while the third is studying Chinese in Shanghai. All are a long way from the corn farm in Milton, or their father's birthplace near Sarnia, Ontario.

Chris Hadfield joined the Air Force as a means to an astronautical end, but getting there, and being there, meant flying planes, lots of them, in lots of places. He had his glider pilot's licence at fifteen, his powered plane licence at sixteen, and has since flown over seventy different kinds of aircraft. He flew trainers at Portage la Prairie, jets at Moose Jaw, and F-18s at Cold Lake and Bagotville. Then he turned his sights to test flying in the United States, specifically at the United States Air Force Test Pilot School at Edwards Air Force Base in California. After graduating at the top of his class there, he found himself as an exchange officer at the famed U.S. Naval Air Test Center in Patuxent River, Maryland, where doing test flights and staying alive were not necessarily compatible.

Far too often, pilots stationed at Pax River were killed just doing their jobs. They "pushed the envelope" too far on test flights, and the planes they flew crashed and burned. All test pilots saw this happen to their friends, yet they made themselves continue in the dangerous milieu in which they operated. The test flight code decreed "that a man should have the ability to go up in a hurtling piece of machinery and put his hide on the line and then have the moxie, the reflexes, the experience, the coolness, to pull it back in the last yawning moment — and then to go up again the next day, and the next day, and every next day, even if the series should prove infinite."[6]

Chris Hadfield had this type of ability, and excelled at flight testing the most advanced aircraft around. While there was always the heartache and the fear, he learned to cope with those as well. "I just didn't think about fear," he told me. "I accepted long ago that I was going to die. I've had lots of friends die as test pilots and fighter pilots. About one friend a year. The death of my first good friend who died in an F-18 really hit me hard. It took me a while to get over that. But when the second one died, and the third one died, and the fourth one died, and the fifth, I started to learn that I could not live with regrets. Sometimes, life is brutally short. That is a fact. So — get things done. Some things are worth risking more for than others. Even if you stay home and hide under a pillow — you're still going to die eventually. So — get something accomplished while you can."[7]

And the young man who ultimately left the Air Force as a full Colonel again became the best of the best. Around the time that the movie *Top Gun* was extolling American bravado in the skies, the top test pilot in the United States was actually a Canadian. And even though Chris Hadfield was named U.S. Navy Test Pilot of the Year in 1992, the achievement is not even alluded to in his official biography as released by NASA.

By the beginning of September 1994, Hadfield had completed his Mission Specialist training and had been assigned to STS-74. However, it would be another year before the flight itself came about. In the interim, Dave Williams, Bob Thirsk, and Marc Garneau were all subjects of NASA reassignments. Williams was named Canada's third Mission Specialist, Thirsk was assigned to his first flight, and Garneau, who by this time had completed his Mission Specialist training, was told he would be making his second flight on the shuttle *Endeavour*.

In the years when Chris Hadfield was going to school, and then pursuing his career in the Air Force, fledgling space exploratory projects were being advanced by both the United States and the Soviet Union.

In those years, many satellites were launched, as were human beings in rocket ships from both nations. But fairly early on, the idea of having some kind of space station had intrigued scientists and others. The soul-searching, dreaming, and invention culminated in the launch, by Russia, of the world's first space station on April 19, 1971.

The thing was called Salyut 1 and, in all, seven such vehicles were sent aloft. "Two of these, Salyuts 3 and 5, were military space stations, and were equipped with high resolution cameras to gather reconnaissance data from orbit."[8] This was, of course, before the collapse of the Berlin Wall, and the Americans and Soviets were highly suspicious of one another. The various Salyuts did accomplish the primary purpose of their existence though, and while reconnaissance was integral to their being aloft, so was the search for knowledge about space. To achieve that end, the vehicles, which would accommodate three cosmonauts, enabled long-term habitation. The space record of the time was set by Valery Ryumin, who spent two periods of time in orbit, for a total of 360 days aloft. The last Salyut was abandoned in 1986.

While all of that was transpiring, and even though it was acknowledged that the Soviets were far ahead in long duration space habitation, the United States was not exactly idle. Two years after Salyut 1 was in orbit, the Americans launched their first space station. Skylab had a crew of three, and while in orbit was inhabited by three separate trios of astronauts. These missions "gave NASA its first understanding of the medical effects of long-term duration space flight; they also produced new data on the sun, the earth, and on the behavior of materials in weightlessness."[9] Skylab ultimately burned up during atmospheric re-entry on July 11, 1979. Another was built, but never launched.

The next step in the ongoing effort to live in space came about when the Russians launched a much larger, more efficient, and, for the time, futuristic space station called Mir. The structure was modular, deliberately designed so that it could be enlarged over time, in a similar way to additions being added to a house on Earth. There were six separate docking ports, where spacecraft from Earth could attach for varying lengths of time, while supplies were delivered and crews interchanged. The station was launched in 1986, and over time was visited by astronauts and cosmonauts from twelve nations; this rotation sped-up following the official dissolution of the Soviet Union on December 25, 1991. Over the years there was a serious fire and other problems with Mir. By 2001, the thing was in bad shape. It disintegrated as it re-entered the atmosphere in March of that year. However, at the time when Chris Hadfield went aloft on STS-74 in November, 1995, it was still functioning.

The Hadfield launch was delayed, but successful. Its main purpose was linking up with Mir and attaching a five-ton Russian docking module. The mission ultimately resulted in Hadfield being the only Canadian ever to visit that Russian space station.

As always, the crowds at Kennedy were large when the mission was about to get underway. They stood and watched in the coldness of the November 13 morning, as the shuttle cleared the pad at 9:30 a.m. The flight had been scheduled to go the previous day, but inclement weather necessitated the postponement. Now, with a lot of luck, the tiny, ten minute launch window remained open, the skies remained clear, and liftoff went without incident. The time slot for departure was narrow, primarily because *Atlantis* had to rendezvous with the already orbiting Mir.

In going on this flight, Chris Hadfield became the first Canadian Mission Specialist to fly, and was, in fact, the senior member of the three men with the same qualification who were on board. He had looked forward to this moment for a long time, and could not have been happier when it came. That feeling was shared by his family as they watched the ten-year-old spacecraft roar out of sight. Helene and the three Hadfield youngsters were all there, and their ages at the time — twelve, ten, and eight — meant that they were all old enough to remember the fascination and exhilaration of the moment. But while they were thrilled to be able to watch this launch, a long-promised visit to Disney World left them with a feeling of anticipation that was almost as profound. Their mother later talked of her feelings at launch time: "I was in tears," Helene admitted. "I wasn't nervous; I didn't have anything in the pit of my stomach. I was just yelling, go, go, go." She said she put some perspective on what she had just seen: "He's done it: it's done," she explained. Of the shuttle mission, and the drive over to Disney World, she added: "He gets his trip. We get ours."[10] By then, of course, the shuttle had already circled the Earth and the work on board had begun.

Atlantis was "NASA's fourth space-rated Space Shuttle orbiter," and over time, it "pioneered the Shuttle-Mir missions, flying the first seven missions to dock with the Russian space station," and "when linked, *Atlantis* and Mir formed the largest spacecraft in orbit at the time."[11] Since then, the International Space Station has far supplanted any previous structure in space in size.

But the International Space Station was still far in the future as STS-74 continued its mission. On board the flight that morning were five astronauts in all, the three Mission Specialists, the Commander, and the Pilot. Awaiting them on Mir were two Russians and a German, the latter representing the European Space Agency.

From the time they entered orbit, until the shuttle re-entry eight days later, the entire crew on board *Atlantis* worked flat out to fulfill the purposes of their mission. Hadfield did his share, and more, and in doing so became the first Canadian to operate the Canadarm. He also opened the doors of the payload bay shortly after the spaceship entered orbit. It was, however, his skills in maneuvering the arm and lifting the docking tunnel out of the bay that proved to be his most taxing achievement. The docking tunnel was large: 4.5 metres long and weighing 4,080 kilograms when on Earth. Lifting the thing from its resting place took time, precision, and a coolness that few humans possess. However, Hadfield did it with a precision that was virtually exact.

The shuttle Commander, Ken Cameron, a seasoned pro in the left seat up front, gradually eased *Atlantis* towards the rendezvous with Mir — like two tiny dots merging in the infinity of space. This operation also took time, uncommon skills, and constant attention to precise placement. Small firings of shuttle engines accomplished the task, and gradually, carefully, and finally, the two spacecraft came together as they circled the Earth at more than 17,000 miles per hour.

In a news conference from space early on in the flight, reporters asked Chris Hadfield about the attachment of the docking tunnel to Mir. "We're doing something no one's ever tried before, trying to build this thing and install it," he explained. "It's definitely a challenge." The Canadian went on to say how thrilled he was to be part of the mission, and to be in space for the first time. "This is a tremendous experience for me," he said. "I've seen all the pictures, but there's nothing like being here. This is a magnificent experience."[12]

The coming together, and then the installation of the docking tunnel on Mir, proceeded without a hitch. Because of protracting solar panels on the Russian craft, easing the tunnel onto it was a slow, taxing maneuver. Once accomplished, however, both spaceship crews were able to exhale again. Then the *Atlantis* fliers prepared to move through the tunnel to visit the three men they had come so far to see.

The men on both spaceships knew each other. In fact, before the cosmonauts on Mir began their flight three months earlier, both crews had trained together for about five weeks. For that reason, the reunion was a happy one, and the newcomers were the first faces the Mir crew had seen since their six month mission began. To a degree the visit swept away what some have called spatial cabin fever. No wonder the arrival of *Atlantis* was a welcome one.

The docking itself was an extremely dangerous undertaking. If the spaceships had collided and one of Mir's solar panels had broken off, or been seriously damaged, the electricity it produced for the Soviet craft would not have been available. A worst case scenario could have resulted in the abandonment of Mir, or even the depressurization of both space-ships. In that eventuality, all eight men would have died. No one wanted a collision under any circumstances.

However, as soon as the docking occurred, the significance of the move elated everyone, from former test pilot Cameron, to Flight Director Bill Reeves back at Mission Control, to the young Canadian on his first mission aloft. "It was an historic day for each of us," Hadfield explained during a press conference from space. "We're all just so happy to be a part of it, and we're happy everything worked out all right."[13]

Once the docking procedures were completed and the precision of the link-up had been checked and re-checked, the astronauts from *Atlantis* made their way through the newly-installed, orange docking tunnel to Mir. There were bear hugs, back slapping, lots of laughter, and shouted greetings in an excited mixture of Russian and English as the eight space pioneers met. Then flags of four nations: Canada, Russia, Germany, and the United States were put up on the walls of Mir, symbolizing the nationalities of those present. Ken Cameron handed Mir Commander Yuri Gidzenko three long-stemmed peach-coloured carnations and a gift-wrapped box of chocolates. These traditional Russian house warming gifts elicited more laughter and pretend-attempts by each cosmonaut to wear his flower with pride.

The party-like atmosphere continued for a short time, but soon the realization that work had to be done brought everyone back to reality. Materials for several science experiments that were to be done on Mir were brought in from *Atlantis*. So were food, water, and other supplies that the men would need in the coming months. Chris Hadfield then came through the tunnel with a collapsible guitar, which he gave to the cosmonauts. He told reporters later that when the multinational group trained together, they often played guitars because music is truly an international language. Before leaving Mir, Hadfield and German cosmonaut Thomas Reiter played a duet together, with Reiter strumming away on a beaten-up relic that had been a fixture on the Russian spaceship. Hadfield's gift guitar was a replacement for the half-broken instrument that Reiter used.

Once the delivery of supplies was finished, and the necessary technical matters had been looked after, the crews of both ships said their goodbyes. In due course, *Atlantis* moved gingerly away from the docking port on Mir and headed for Earth and home. As Ken Cameron brought his craft to a 226 mile per hour touchdown at Kennedy, Mission Control greeted him in Russian, and complimented him and the crew on a great mission. Cameron replied in Russian, but added in English that they were all happy to be back. The last shuttle flight of 1995 was over, and STS-74 was now history.

9

Fear Comes in the Months Before

Bob Thirsk was in grade three in Powell River, British Columbia, when John Glenn circled the globe in Friendship 7. That mission, coming less than a year after Alan Shepard first flew, finally established the United States as a bona fide contender in the space race. At the time, little eight-year-old Thirsk had a somewhat limited understanding of what the space race actually meant, but when he saw the pictures of Glenn and the capsule, he was completely entranced. He could barely believe that a human being could ride around the world in such a tiny spaceship, and then come back to talk about doing so.

"Glenn was doing something very few people had ever done," Thirsk explains, "and he was flying in such a hazardous environment. I was amazed by the whole thing."[1] The Glenn story had an immediate impact on young Thirsk. He began to see a much greater value in mathematics and science, and even at the elementary school level knew that those subjects would be important in John Glenn's world. "I went to the library, devoured flight magazines, and things like National Geographic — and anything else I could get my hands on so I could find out more about space programs," said Thirsk. "[This]was years before we had any astronauts in Canada. But even then, I started collecting pictures of Russian cosmonauts and American astronauts, in much the same was as other kids wanted hockey cards. I guess you could say I was hooked," he says today.

At the time, Bob Thirsk's father was in sales at a hardware supply company. Each time he got promoted, the family moved. And there were lots of promotions. That was why, when Thirsk flew on *Columbia* in late spring 1996, several Canadian towns claimed him as their own; including the previously-mentioned Powell River, Port Coquitlam, North Surrey, Kelowna, Calgary, Winnipeg, and Montreal. Thirsk was in Montreal, and a graduate in medicine from McGill, when he became one of the country's original astronauts back in 1983. A little over three months later, he

Bob Thirsk flew on board the shuttle *Columbia* in 1996. In 2009 he became the first Canadian to live and work on the International Space Station.

Courtesy Canadian Space Agency

was named backup for Marc Garneau on his first mission, but the patient and persistent Thirsk had to wait twelve years before he finally went into space himself.

Because Thirsk had an older brother who loved math and science, the shared passion affected both boys. "He was my role model when we were kids," Bob Thirsk remembers, "and to some extent, he was why I started to consider going to university and perhaps getting a degree in engineering. I guess I thought that would be cool." Whether it was cool or not, engineering was important in the life of the man who would soon become his country's first ever long-term resident of the International Space Station. Once he got the opportunity, Thirsk thrived at university and obtained two degrees in engineering.

"In my third and fourth years, we started to do more independent research, and I found myself choosing projects that had medical aspects to them. From my reading in physiology and anatomy, I came to the understanding that the human body is the most complex and marvelous piece of engineering there is. That was why I really became attracted to bio-medical engineering."

One day, towards the end of his time at the University of Calgary, one of his professors took Thirsk aside and suggested that if he really wanted to have a career in biomedical engineering, he might be wise to get a medical degree as well. At that time, the idea had never crossed his mind; he had no family or friends who were doctors, so there had never been any encouragement for him to consider medicine. But the idea had immediate appeal.

"I had to pick up the necessary pre-med courses first, but after that I got accepted to McGill in Montreal," Thirsk recalls. "I was still in Montreal when I heard that the Canadian government was looking for astronauts."

That childhood fascination with astronauts and their exploits in space was about to become an irresistible force; thirty-year-old Doctor Thirsk was Chief Resident at Queen Elizabeth Hospital in Montreal when he stopped by the staff lounge one afternoon, and happened to pick up a newspaper that was lying there. A front page story about Canada's call for astronauts caught his eye. "I got very excited," he says, "because it brought back this dream that I had had for years."

So he applied — along with the several hundred others mentioned earlier, including Marc Garneau, with whom Thirsk would soon work closely.

I asked Bob Thirsk to tell me the reaction to his application and then selection from his family and closest friends. "Everyone was supportive," he told me, "particularly my girlfriend then, my wife now, and my parents. They had never tried to force me to go into any particular job. When I was young, my parents let me know what opportunities were out there, but ultimately, it had to be my choice anyway. Brenda and I have three children ourselves now, and we try to support them in what they want to do. That's the kind of encouragement I got when I sent in that application."

Selection day for astronaut candidates was not easy. Waiting, hoping, filling time, watching the clock, preparing for the best, for the worst, keeping busy, and longing for and dreading that phone call took reserve of many kinds. The day was Saturday, December 3, 1983. "In order to relieve the stress," he explains, "I took Brenda to a movie, because I wanted to get my mind on something else. We had all been told that we would be called; starting at 6 pm that day, but up to that time, there seemed to be a lot of hours. I remember that after the movie I went to the gym to work out, but I was back before six, waiting for the call. It came at 6:05, so I felt pretty good. Brenda was really excited for me. She could see the passion in my eyes and in my heart for this new career change."

Right after the call, "Thirsk announced that they were going to celebrate by having dinner in one of Montreal's exclusive restaurants. It was an indulgence they'd discussed before and rejected as too expensive, but Thirsk was not to be dissuaded, particularly since he had an engagement ring for Brenda in his pocket. He proposed that evening and they were married in early January."[2]

The Canadian requirement that astronauts be bilingual in French and English was not a problem — Thirsk was already equally at home in either when he applied for the job. Later on in his career the study of Russian was another question entirely. "People often ask me what is the most difficult part of astronaut training," he says today. "I sometimes point out to them that the space shuttle is a very complicated space vehicle. The Soyuz is a very complicated space vehicle, and the space station is extremely complicated. There are four million lines of computer code that control the space station — so learning to operate with those three vehicles has been a challenge for me — but it's not as challenging as learning another language, and particularly Russian."

In the most recent part of his career, Thirsk has devoted hours and hours to learning, and then mastering Russian. "It's been a tough struggle," he admits. "I have enough Russian to go downtown in Moscow and shop, or have a conversation, but to function well in a simulation, or in the heat of an emergency, I still don't feel completely confident. I have had two sessions of Russian immersion, where I lived with a Moscow family, and attended classes every day, but I'm still learning."

Back when he first became an astronaut, Thirsk's learning curve was decidedly steep as well. This was even more pronounced because of his selection as Garneau's backup. The two were quickly sent to Houston, and like novice swimmers being tossed into the deep end of a pool, had to fight, flounder, and focus their every waking moment in preparation for the flight. While Thirsk had to be as knowledgeable as possible for the mission, he knew all along that he was never going to go on it. However, once it was over and successful, Thirsk was expected to help with the publicity barrage that Canada's first man in space endured. He and Garneau were interviewed hundreds of times, spoke to crowds almost as often, and trekked from coast to coast to coast because everyone in the country wanted to touch our new spaceman.

Gradually, the media frenzy began to die down and life returned to a semblance of normality. By this time, even though he had yet to fly, Bob Thirsk had not only endured, but thrived on the work expected of astronauts. "The first couple of years, we are in basic training," he explains. "All new astronauts follow the same course, basic instruction on space-related subjects: astro-physics, planetary geology, terrestrial geology, oceanography among them. Then you learn about individual systems that make up space vehicles, such as the space shuttle and the International Space Station — the power control system, the electrical system, communications system and so on. The idea is to bring everyone's knowledge up to a basic level. Astronaut candidates could be engineers, medical doctors, scientists, and more recently, educators as well. All of these people come from different backgrounds, so the basic training is to bring everyone up to a common level of understanding. A Pilot is going to understand how to insert an IV; a teacher is going to learn how to fly a jet aircraft. Then, after basic training is completed, training still continues, but at a slower pace — on say, the skill areas EVA (Extravehicular Activity or spacewalking), Robotics, and Rendezvous. There is time then for the astronaut to provide services, for the CSA or for NASA, for example, because at this point the individual has several skills. We call these collateral duties, and CAPCOM (Capsule Communicator) is one of these. I was a CAPCOM in the space station control room, and I found the job particularly satisfying."

The job of CAPCOM is done by astronauts who are specifically selected and trained to communicate with the shuttle when it is on orbit, or with the space station. It is an important and taxing job, as the individual doing it is the main link with the crew on either space vehicle. Other Canadian astronauts have filled the role over the years.

Bob Thirsk flew on STS-78, at the time referred to as the Life and Microgravity Spacelab Mission. For him, it was undoubtedly a highlight in a series of highlights in his astronaut career. The mission lasted slightly less than seventeen days, from June 20 to July 7, 1996, and forty-three scientific experiments were conducted on it. *Columbia* carried a crew of seven, one of whom was a woman. Included in the crew makeup were two physicians, a veterinarian, and a physicist, and they hailed from three nations in all. Thirsk was the fifth Canadian to fly.

Departure day weather was mixed in central Florida. Clouds and reported thunderstorms over Atlantic coastal regions caused some concern for NASA, but proved not to be bothersome when it came time for the early morning launch, which went well.

"My family was there," said Thirsk, "including our youngest, who was only five weeks old at the time. I'm told that he slept through the whole thing. As you know, the launch is quite dramatic, and there is some risk involved in it, so there was some relief when it was all over. One of the things I did, I wrote, prior to the launch, letters to each of my family members, and they were not given the letters until we were on orbit. I was letting them know that they were not alone watching the launch; I was right there on the building rooftop, watching with them."[3]

I asked Thirsk about the fear factor involved in his job, particularly during launch and subsequent re-entry. His response was detailed and illuminating: "I think that astronauts, their spouses and family members deal with the fear weeks and months before the launch. There is a chance that it will end in disaster, and those chances are not insignificant. But our training makes us very focused on our launch day duties, and therefore, ninety percent of the time I was in my seat on the shuttle on launch day, I was focused on the duties I had to perform if things went wrong. So fear's not an issue on the pad, but it is something you consider in the months prior to that. You have to ask yourself: 'Is this something I really want to do? Do I want to subject my family to this kind of risk? Do the benefits of what I am about to do outweigh the very small chance of an accident that could injure or kill me?' Because we are strapped in the shuttle for two and a half hours before liftoff, there were moments when I thought about the crewmembers of *Challenger. Columbia* had not happened then. But I thought back to the *Challenger* astronauts, some of whom I knew personally. But those were just fleeting thoughts. Our training is excellent, and we can focus on the tasks ahead." Even though the training is first rate, Thirsk admits that sometimes a little extra help might be in order. He recalls saying a quick prayer in the moments just before blastoff.

His answer led me to ask specifically about his internal response to being in space. It is well-known that the experience affects some astronauts in ways they may not have necessarily anticipated — despite their training and extensive preparations for whatever mission, or missions, they flew. Thirsk feels he is essentially the same person he was before his flight, but that he has more credibility in the field because he has flown.

There have been instances, however, where those who flew did not seem to be the same afterwards. This was particularly true for the moonwalkers in the Apollo program when they returned to Earth. Some drifted from job to job, others became alcoholics, many suffered marital breakdowns, and a few actually mellowed. But one, Charlie Duke, "Came home and imploded … He'd lost his moorings and been unable to settle; had terrorized his children and tormented his wife."[4] Ultimately, Duke found peace through religion. Buzz Aldrin, the second man on the moon, spent years in an alcoholic haze. Jim Irwin was sure that God was speaking to him from a peak in Italy's Apennine Mountains. John Young turned on NASA, harshly criticizing the organization that gave him fame, while Neil Armstrong became, and still is, a highly reclusive global hero. "Where do you go after you've been to the moon?"[5]

Bob Thirsk says the flight he was on did have an impact. "Until you have flown in space," he explains, "you cannot understand the living and working constraints that a microgravity environment imposes. So I did come back from that flight knowing I had a unique experience, and I find now that the contributions I can make to the Canadian Space Program in terms of mission planning, hardware, and software design are more credible than before I flew. I just have a better understanding of how you need to operate in space — so from an operational point of view, I feel I have more credibility and experience. On a personal level though, I think I am still the same person.

"There is a psychological term that pertains to some astronauts, and the term is called the Overview Effect. This is an emotional or psychological personality change that some astronauts undergo. There is an emotional or spiritual awakening that occurs in their mind, in their soul during a space flight, so much so that they become dramatically different people. That did not happen to me.

"When I was in space," Thirsk continues, "certainly I looked out the window. I saw the beauty of the planet down below. I saw it with clarity; I saw heavenly bodies out there and I witnessed sixteen sunsets and sixteen sunrises a day, and I contemplated the meaning, the majesty of life on this precious but fragile planet, but I don't think I changed in a big way. I think I am still the same person I was fifteen or twenty years ago. The only caveat now though, is that I have become more of an environmentalist. I was an environmentalist prior to my flight, but when you are on orbit, and have that orbital vantage point, you have the opportunity to witness some of the tragedies and the mistreatments that humans have imposed on some regions of the world; you cannot come home without a stronger environmental spirit. All astronauts come home with an understanding that this planet is a lot more fragile than they might have thought it was."

In the years since he flew, Bob Thirsk has been a busy man, and he has never doubted the appeal of his profession. "I spend ninety percent of my time doing operational, scientific, or technical tasks," he told me, "and all these tasks are at the cutting edge, the leading edge, of what we do scientifically, technically, medically. So, for example, with every new shuttle flight to the International Space Station, we are doing something we have never done before — in terms of robotics, in terms of EVA, assembly techniques. The media do not pick that up; they don't inform the public. The media cover the launch and landing, but those are not the interesting things. The really interesting things are the things which we are doing on orbit, which are mind-boggling. Because the challenge is still there, I am not bored with what I am doing. Because I enjoy challenge, I am not frustrated; I am not bored."

Bob Thirsk spends between five and ten percent of his time doing public relations, talking to students in schools, and dealing with the media. The work with young people is particularly gratifying he says, because it is his way of touching base with youngsters who have their careers ahead of them. He cites instances of senior students approaching him after an address, and mentioning that because they heard him speak years earlier at their elementary school, they decided to work harder and focus their attention on getting good marks and graduating. That type of reaction, he feels, goes a long way towards justifying the human space flight program in Canada.

Not only Thirsk, but all of our astronauts have become role models for young people. In research for this book, I witnessed that first hand. All that any member of our astronaut corps has to do is walk into a school auditorium, and the young people present look at the man or woman in the blue flight suit with awe. When the astronaut speaks, and later signs autographs, the reaction is overwhelming. This person has been up in space. What this person says is important. This person is my hero! So when an astronaut urges audience members to stay in school, work hard, and graduate, the message is taken to heart.

But our astronauts also have heroes themselves; individuals they looked up to along the way as they established their credentials in the space program. Bob Thirsk readily admits he has a role model, and was even able to bring his admiration for that person to his job. In fact, Thirsk's hero even had a tangential, but creative role during the flight of STS-78.

One of the finest hockey players to ever lace skates is the legendary Bobby Orr, the all-star defenceman who spent most of his career with the Boston Bruins. Thirsk plays and coaches hockey. He credits Orr with being an inspiration for him, not only in that sport, but in life. The two have met, and Thirsk calls the Hockey Hall of Fame member a "quintessential human being. He's very humble. He's a good role model, and to me, he was the best hockey player who ever played the game."[6]

Bobby Orr played hockey for several years, winning both trophies and accolades for doing so, but is perhaps remembered most vividly for a memorable Stanley Cup final series goal he scored for Boston in the spring of 1970. A famous photograph of the feat is "firmly set in the minds of Bostonians ... of an exuberant Orr flying through the air past St. Louis defenceman Noel Picard after the Beantown hero had fired the cup-winning sudden death goal past goalie Glenn Hall."[7] The photo has been reproduced countless times, and prior to his shuttle flight in 1996, Thirsk decided to try to replicate the feat in space.

He contacted the retired hockey player, told him what he had in mind, and awaited the reaction. A few days later, Orr sent him not only a game jersey, but the Stanley Cup ring he had won in that Boston Bruins–St. Louis Blues series in 1970.

Thirsk admits that he was in awe of Orr for years, but when he received the package containing the ring and sweater, was almost at a loss for words. He then had to get NASA's approval to take both with him on the shuttle. Even before he got that okay, he had pretty well decided how he intended to use them. When an appropriate opportunity came during the mission, Thirsk donned the sweater, slipped the ring on his finger, and as a crew member snapped pictures, pretended to fly like Orr, immediately after scoring his most spectacular goal. In Thirsk's case, however, he was moving, weightless and headlong the length of Spacelab, wearing the sweater and ring, shorts, and white ankle socks instead of skates. The photograph clearly indicates the delight on his face as he made the move.

But there was much more to STS-78 than the antics with the Orr artifacts. During the seven million miles of the flight, the crew on board performed a host of experiments, manoeuvres, and studies. Bob Thirsk was a party to many of them — some more significant than others. On

Courtesy Bob Thirsk

Canadian astronaut Bob Thirsk always idolized Bobby Orr, the legendary defenceman for the Boston Bruins. In this photo, Thirsk is wearing Orr's sweater and Stanley Cup ring, as he imitates Orr's flying across the ice after scoring his team's winning goal in the 1970 NHL final.

Canada Day, July 1, 1996, for example, after the crew had been wakened by the playing of "O Canada" from Houston, several of the astronauts conducted experiments to assess the effects of microgravity on the pulmonary system. These studies were repeated on subsequent days, and the results were tabulated and compared each time they were done. During the tests, three electrodes monitored heart activity, while a vest-like contraption, called a Respitrace Suit, kept track of rib cage and chest motions. The suit had to be modified as the flight continued because, as always, the spines of those taking part grew longer in space.

Other experiments were conducted by Thirsk and his colleagues that involved hand grip strength tests. These were done in an attempt to ascertain the degree by which space flight effects energy, muscle strength, and, ultimately, fatigue. Thirsk and another crew member, American Mission Specialist Doctor Richard Linnehan, were also the guinea pigs in experiments done to test eye, head, and body coordination during each day on orbit. The two wore head and torso sensors for these tasks; it was felt that the results of the tests would "help researchers identify causes of motion sickness during spaceflight, develop counter measures and lead to practical ways on Earth to avoid motion sickness in cars, boats, or aircraft."[8] Such evaluation is ongoing.

As *Columbia*, and those on board, continued their odyssey, not only family members, but countless others were interested in the mission, and followed it in different ways. The Canadian media provided coverage on both radio and television, and in the major print outlets. Some reports were handled by the *Canadian Press*, which often resulted in the same stories being reproduced widely. Sometimes local papers highlighted reactions that reflected the views of specific earthbound audiences. In Vancouver, for example, the *Sun* sent a reporter to an area school to see how students there were taking the whole thing.

At Hjorth Road in Surrey, where Bob Thirsk spent his fifth grade, the youngsters the reporter encountered held widely varying perspectives on the flight. All were tremendously excited as they sat on the gymnasium floor to watch a videotape of the *Columbia* launch. The fact that it had taken place hours earlier did not dampen the enthusiasm. The entire student population was present, and according to the reporter, the tension in the room was palpable because many of those present were not aware of the time-delay, but undoubtedly would have reacted with enthusiasm anyway. "As a radio voice counted down to zero," noted the reporter, "small hands stopped fidgeting and bodies leaned forward. When the craft lifted off, the gymnasium erupted in cheers and applause."[9] The reaction was much the same at the Cape, in Houston, and wherever television viewers looked upon the launch.

Because this flight of *Columbia* was long, and the various experiments critical to its success, it was satisfying to the astronauts on board that they were able to complete so many of their intended objectives. Doing so meant hard work, often too little rest, and acclimatizing to microgravity and functioning in it. The rest factor was, and always is, somewhat problematic in space flight. In fact, Bob Thirsk said he always found himself waking up tired each day, in spite of what might be viewed as ideal sleeping conditions.

"On orbit, we sleep in sleeping bags," he explains, "and sleeping on orbit is the most comfortable possible. There are no pressure points. When you completely relax, your body goes into something that resembles the fetal position. There is no mattress pressing against you. It's like drown proofing for a swimmer. I would fall asleep within three to five minutes, and would sleep through the night. But — I would wake up tired.

"On our flight, among the other medical experiments was one on sleep study that we did for the University of Pittsburgh Medical School. Five of us participated. We put all kinds of sensors on our bodies, and our brain waves were monitored. There are four different stages of sleep, and the last is the most restful. You cycle through the stages, every hour and a half or so. But in space, we determined that we never experienced the most restful stage. So that was why we woke up tired.

"But the tiredness vanished quickly because we had so many exciting things to do. That was okay for seventeen days, but you could never continue like that for six months in space." Bob Thirsk is now in space, and will be for several months. When that time comes we wish him all the best — and pleasant dreams.

The sleep experiments were not just designed to help those who spend time in space, where "sunrises and sunsets occur every hour and a half as the shuttle circles the earth. Their experiments could benefit shift workers on earth and people suffering from sleep deprivation and jet lag."[10] In fact, a lot of advances in such areas have been made, but only so much was possible on STS-78. All too soon for those on board, the mission ended, and the shuttle was lined up for landing at Kennedy. The tires of the spaceship touched the concrete surface of Runway 33, coasted 9,290 feet, and came to a complete stop. Then the *Columbia* would be inspected for safety concerns, refurbished, and readied to go again — four months later.

10

You're Going into Space, or You're Going to Blow Up

The Engineering Professor at the University of Western Ontario in London leans back in his chair, and rather reluctantly begins to talk about himself. He tells you he was born in Iceland, and while he only lived there for the first seven years of his life, explains how a small incident when he was six had a lasting impact on so much of what he has done ever since. "A cousin who was about fourteen came to visit our home in Reykjavik," the professor explains. "I didn't really know him at the time, but when he showed me some pencil drawings that he had done, I became quite interested in him and in what he had drawn. I did not know what the things were though, so I had to ask. He told me they were airplanes."[1]

At this point in his life, Bjarni Tryggvason had never seen an airplane, or even a picture of one. Forty-six years later, he became the only person from Iceland to circle the Earth on a space shuttle.

The intervening years were long and circuitous, but the images of those airplane sketches were never far from Tryggvason's mind. In fact, the cousin gave him one to keep, and he treasured it for a long, long time. A year or so after getting the drawing, the young Icelander not only saw his first plane, he flew in it. That was because the Tryggvason family was leaving Iceland and coming to live in Canada; in Nova Scotia.

Because there was no direct air connection between Reykjavik and Halifax then, or even today, the move entailed flying to New York, and from there on to Nova Scotia, by way of Montreal. All three of those flights left young Bjarni with an unquenchable love for planes, for flying, and for the aura of adventure that the machines evoked. Even today, as a member of Canada's first astronaut group, his eyes shine when he talks about flying — in any type of aircraft. But it was that journey to Canada that instilled a love of the air in the man who today has over five thousand hours in his log books.

"We flew to New York on a DC-6 or something," said Tryggvason, "and I spent the entire time glued to a window, looking down at the Earth

Bjarni Tryggvason, who flew on STS-85, was born in Iceland and still is the only astronaut from that country. He became a Canadian when he was a youngster.

Courtesy Canadian Space Agency

and ocean, and on the clouds in particular. I just couldn't figure them out. I didn't know if they were real clouds or fog, but I was intrigued. Then, when we landed in New York, I saw something I will never forget."

As anyone middle-aged or older will recall, at one time airports and the planes that landed and left them were not cocooned in the veils of security that we encounter today. Being able to walk out onto the tarmac to greet arriving loved ones was commonplace; even going out just to look at planes was permitted. That was the way it was at the time. However, the day the Tryggvason family landed in New York, one of the first things they were told was that no one was allowed to go out onto the tarmac. That was because a jet plane was parked out there!

"Now, I had never seen a jet plane before," Bjarni Tryggvason says today, "but it really made an impression on me. First of all, there were no propellers, and I thought that was amazing. And this jet was so small, and it had such an exciting shape. I can still remember both those things. In fact, I can still see the plane parked there. Then I asked, or probably my father asked, what kind of plane it was. It turned out to be a T-33. Ironically, years later, I flew T-33s when I worked at the National Research Council in Ottawa."

The Tryggvason family moved on to Nova Scotia, and remained there for two years before going to British Columbia, where the future astronaut grew up and continued his education. Today, he cites both provinces as wonderful places to be young. He regales you with stories of his formative years, of games, fishing, swimming, sports of all kinds, exploring, and being happy and carefree in days that seemed to last forever. In British Columbia, the family lived in Kitimat for four years. When every street and backyard in the town had been checked out, he and his friends would hitchhike elsewhere, often as far as forty miles from home. There was no danger in doing this, he explains, or if there was, it was never really considered. "We lived a block from the forest, so we spent a lot of time running around in it. When we were not in the forest, we were on our bikes, going all over the place."

Then, one day a marvelous thing happened. "When I was in grade six, Sputnik was launched — and that made a huge impact on me. It was the same kind of impact that the drawing of the aircraft had when I was much younger. When we first heard about Sputnik, everyone was doing normal things, going to school, going to work, whatever. But when we found out about that satellite, it made me realize for the first time that there were people out there thinking of grander things, and doing amazing things that the rest of us knew nothing about.

"It was the launch of Sputnik that gave me a kind of motto or purpose in life that I never had up until that time. It impressed me to the point that I decided then and there that I was never going to let people around me or the things around me be a limit to what I could do. That has been my intent ever since."[2]

In 1959, the Tryggvason family moved to Vancouver, where Bjarni's father became involved in the fishing industry. The son continued his education during the week, but on weekends, and some evenings, he spent as much time as he could in the Air Force Reserve. There, he learned all he could about planes; everything except flying them, although several pilots who also worked there liked the young Icelander, and took him up on occasion. In the two summers he spent with the reserve, he did everything from aircraft repair to participating in search and rescue operations at the Air Force base at Comox, on Vancouver Island. And — he recalls — got paid for doing what he loved.

At one point during his teenaged years the young cadet thought he might join the Air Force itself, largely because he wanted to fly jets. In time, he did fly jets, but did not have to go the military route to do so. One thing that's quickly apparent when talking to the man, Tryggvason embodies a mind and spirit that are obviously far too freewheeling to conform to a military environment. "And I knew this," he explains. "I knew myself well enough to realized that army or air force life were not for me. I knew I was not in tune with the kind of discipline needed in the military. I prefer to think my way through things and do them because they make sense, not just because someone tells me that something has always been done that way."

Nevertheless, Tryggvason still wanted to become a pilot, but he also intended to go to university, in case a career in flying didn't pan out.

"So, one day," he explains, "before I decided to go to university, I was sitting at the kitchen table at home, talking to my mother and my brother, and she asked me what I was going to do. I told her I wanted to be a pilot. So she said, well, you better get going then. That was it. We finished our coffee; I got in the car, and started knocking on doors, asking how one went about becoming a pilot. I ended up at Langley, [B.C.] and told them what I wanted to do. I said I wanted to become a commercial pilot, but they told me I had to become a private pilot first. So I said okay. Then they asked me when I wanted to start. I said, right now. I soloed one week later, largely because having been in the reserve I knew how a plane worked, where the instruments were and so on.

"At the time, you had to have about thirty-five hours to get your licence, but one day, when I had about twenty, the instructor said to me: 'We don't know what to do with you. You have completed everything, but we can't give you a licence because you are fifteen hours short of the time required.' So I asked if I could go and learn to fly float planes in order to get the rest of the hours. So I did. I got the float plane experience, and the hours as part of the private licence. I also went on to get my commercial and did a lot of float plane work, along with aerobatic stuff. Then I thought I had better get to university."

Bjarni Tryggvason went to the University of British Columbia where he earned a degree in engineering physics. He was involved in postgraduate studies at the University of Western Ontario when he was offered a job at the National Research Council, in Ottawa. When the first advertisements for astronauts came out, he was in the Capital. By that time, he was flying those T-33 jets.

"When I first looked at the advertisements for astronauts," he says, "I thought to myself: 'My job has come about.' When I applied, they had really no idea what they were looking for, and nor did I. But I got through the first selection gate, along with 1,800 others. Then they sent out more information, and when I looked at it, I decided I was going to make the next round. Sixty-eight of us were picked. That was when I really started to figure out what they seemed to want, and I became determined to get there. Finally, they put nineteen of us on a short list, brought us all to Ottawa, and we met each other, as well as the selection people. At that point, I decided there were no problems here. I looked over the group and picked five of the final six. I was one of them."

For everyone except for Marc Garneau, the period between the selection of the first astronauts and the time of their flight was both long and complicated. NASA was plagued with everything from money problems, to uncertainty, to the terrible loss of *Challenger* and its entire crew. In Canada there was much speculation as to how many Canadians would even be assigned a mission. Initially, there was supposed to be only two. However, because of the extensive buildup to the astronaut selection, and the resultant publicity, informed observers thought there would have to be more than that. In time, a third place opened up, and the six hopefuls from Canada looked upon three flying and three backup. In that way, all would be involved. Ultimately, however, as we all know, five of the first six flew, Tryggvason among them.

Thirteen months passed between STS-78, with Bob Thirsk on board, and STS-85, Tryggvason's mission. During that time, and earlier, he and several other Canadian scientists scrambled to assemble suitable experiments for the flight. When his time came, Tryggvason went as a Payload Specialist; his primary role was performing fluid science experiments in space. This work was vital in the examination of various liquids and their sensitivity to spacecraft vibration. Ultimately, his work contributed to the understanding of the effect that vibrations have on experiments being performed today on board the International Space Station. The background of what is being done now has a large Canadian component, and men such as Tryggvason helped bring that about.

The weeks leading up to the Tryggvason flight became a whirlwind of activity for him. Final preparation for the experiments he would do took time. So did crew training in Houston, the requirements to maintain a high level of physical fitness, and the never-ending, incessant media requests for stories that they were developing. Rather expectedly, Tryggvason's background became a factor. He had been a Canadian citizen for years, but his Icelandic heritage generated a great deal of interest in the land of his birth. The press there ran pieces about him; everyone on the island knew his name, his picture was circulated widely, and school children called him their hero. There were even rumours that Iceland President Olafur Ragnar Grimsson might go to Florida for the launch. Then the rumours became fact: President Grimsson did go to the Cape for the big day, and he brought about fifty of his countrymen with him — among them, some of Tryggvason's relatives. Other family members came from Canada, of course, many of them from British Columbia. Unfortunately, Bjarni Tryggvason's mother had passed away by that time, and his father was too frail to be present.

The Tryggvasons themselves had split up about five years earlier, but his ex-wife Lilyanna Zmijak, and their son Michael and daughter Lauren were both there. Tryggvason thought of them as his shuttle *Discovery* sat on the pad just before the launch.

"I always remember saying to myself: 'let's get this countdown done, so they can press the button and we can go. Don't give me any of those last minute holds.' I also hoped the people who were watching had a good show to watch — but not a spectacular one. When you are sitting on the pad, you have only two possibilities: you are either going into space, or you're going to blow up."

Tryggvason, who recently flew a replica of the legendary *Silver Dart*, also flies World War II Harvard aircraft, compares a shuttle launch to the formation flying he loves to do whenever he can. "Formation flying demands your undivided attention," he explains, "you do it right, and everything's okay. But if you misjudge by a second and you hit somebody else, you wipe out. If you blow up in the shuttle, or wipe out in formation flying, those are the risks you take."[3]

Then Tryggvason talks about the spectators at a launch: "Those people down there who are watching — especially your family members — they didn't make this decision that you take this risk. They are just there, supporting, watching, and hoping everything's going to be okay. I talked about the risk with my family, and they knew of the risk as well. My kids were about nine and eleven at the time, and they understood the risk too. And even though they were not old enough to remember the *Challenger* accident when it happened, they had seen it replayed on television so often that they knew.

"My kids asked me: 'Why are you doing this? You might blow up and die.' I told them I was willing to take the risk, because it was something I had to do — so they were very supportive. But after I had flown, they told me: 'Okay, you did it once. You don't have to do it again.' And I didn't, although I likely would have had the opportunity. I made the decision not to go again because I wanted to try to be of more help to my kids. I completed the Mission Specialist training, but if I had begun the training for another mission, I would not have been able to see them for far too long." As Bjarni Tryggvason says this, you sense the depth of his intention. He is extremely proud of his son and daughter, and shows pictures of them. Lauren attends Wellesley College, in Boston, and plans on becoming a veterinarian. Michael will soon have his engineering degree — from Western, where his father teaches — and like his father, flies both float planes and other commercial aircraft.

Discovery blasted off on time, and newspapers around the world covered the story. One of them described the launch in this way:

> Bjarni Tryggvason rode a fiery trail of smoke through the Florida sky into the deep-freeze of space yesterday morning, ending a 14-year wait on the ground. Thousands of spectators, including 400 Canadians who traveled to Florida for his launch, cheered as the shock waves raced across the ground after ignition. But the roar of engines drowned out the cheering, firing out 7 million pounds of thrust and lifting *Discovery* through haze and light cloud.[4]

The paper went on to explain that the fourteen year reference was how long Tryggvason waited to fly, and that he was the last of the original six Canadian astronauts to do so.

As the crowd on the ground watched what Tryggvason later learned was "a good show," they screamed themselves hoarse, expressed amazement at the Gatling gun retort of the shuttle engines, and expressed both relief and disappointment that the drama ended so soon. *Discovery* raced out over the Atlantic, was still gaining altitude as it swept past Iceland, and ultimately entered orbit two hundred miles above the Indian Ocean. There, the pull of gravity disappeared, and anything not secured on board began to drift around the decks. Once that happened, the six astronauts on board — five Americans and one Canadian — set about preparing for the work that awaited them.

By the time the spectators at the Cape stopped looking to the heavens, the smoke from the shuttle engines had wafted away, and the reporters present were gathering reactions for the stories they would file. One of the first to comment was Icelandic President Grimsson. He expressed amazement at what he had just seen, and immediately claimed Bjarni Tryggvason as Iceland's "first space voyager." Not far away, friends, family, and countrymen of the crew were equally enthusiastic. NASA said that STS-85 had been "a perfect launch."

During the almost twelve days that followed, the perfect launch became a highly successful flight. Commander Curtis L. Brown, Jr. guided his craft and crew in a professional and businesslike manner. The mission was his fourth, and in ensuing years he flew two others, one of which was also on *Discovery* when astronaut pioneer John Glenn returned to space at the age of seventy-seven.

As STS-85 progressed, so did the work being done by Tryggvason. He was praised for his contribution and for his unwavering attention to the experiments he had been entrusted to do. He brought with him a briefcase-sized instrument called a Microgravity Isolation Mount (MIM) and worked slowly through several math formulas in search of the perfect balance in testing it. The fluid science experiments designed to examine sensitivity to spacecraft vibrations were, as mentioned above, vital in his role.

There were also several other aspects to the mission. One of those involved the release and then retrieval of a small satellite, which gathered data on the Earth's ozone layer. Another involved the testing of the prototype for a new Japanese robot arm. The device was about five feet long and had three fingers. Unfortunately, its performance was somewhat disappointing, but the retrieved satellite captured enough priceless data to keep scientists occupied for months. The crew also observed the Hale-Bopp comet as they sped through space.

The shuttle landed back at Cape Canaveral, where the touchdown was smooth. In the words of a news agency report: "A full moon shone as *Discovery* swooped through a pale sky and rolled down the runway minutes after sunrise yesterday."[5] The landing date was August 19, 1997.

Soon after *Discovery* came to a stop, and the crew on board had been checked by NASA physicians, they posed for group photos, and told the assembled press people how great it was to be home. They were all enthusiastic about the 4.7 million miles they had traveled, and all professed a willingness to do it again if circumstances warranted. But, as indicated before, the *Discovery* mission was Bjarni Tryggvason's first and final flight into space. He continued to be associated with NASA until September 2005, when he was seconded from the Canadian Space Agency to the University of Western Ontario. There, as the school alumni magazine reported, he established "a research facility at Western Engineering, collaborating with faculty members and graduate students in his field, and sharing his phenomenal experiences with our undergraduate students."[6] Today, Tryggvason says he is happy at Western, and remains enthusiastic about his work.

I asked him if his time in space changed him.

"No, I don't think so," he replied, "but it did have one effect that has been lasting. It made me realize what a mess we are making of this

Doctor Bjarni Tryggvason is now a professor in the Department of Materials and Mechanical Engineering at the University of Western Ontario. He also flies WWII vintage Harvard aircraft.

Courtesy John Melady

world. I remember years ago, when I first came here and was flying planes, you could take off from London, and if you climbed high enough, you could see the CN Tower in Toronto. Not any more. On the clearest days there is haze. I remember flying across the prairies years ago, and the sky was crystal clear. You don't get that now.

"So many of the pictures we have all seen of the earth from space show the earth as being clear and beautiful. But take pictures of China, India, Borneo, Indonesia, Africa, and South America, and they are not clear. From space, you see the smog, the smoke, the dust, the haze, the burning of forests, and so on. I remember coming over Borneo, and I was trying to take a picture of it. Well, you could hardly see it. There was all this muddy stuff around the islands, because the soil is being washed into the ocean.

"In many of the talks I have given since my flight, I tend to talk about those things rather than the gee whiz of the flying. We have learned what we are doing to the planet because we have seen the results from space. Talking about that and trying to do something about it is more important to me than talking about the gee whiz part. You know, that talk that our friend Al Gore received the Nobel Prize for was the same talk, using some of the same slides that I have been doing since 1998. I guess it helps if you've been a Vice President."

On that note I said goodbye to the Professor, and as I drove away from the Western campus, thought of all the summer smog alerts in Southern Ontario. Perhaps it is time we really listened to Al Gore — and to Bjarni Tryggvason as well.

11

Under the Ocean and Over the Clouds

On Saturday, March 1, 2008, one of this country's most accomplished space pioneers retired from active astronaut status. In his sixteen years with the Canadian Space Agency, Doctor Dave Williams flew on two shuttle missions, did three spacewalks, spent 687 hours on orbit, and travelled almost twelve million miles in the process. In ceasing active flying he was leaving a career that was outstanding, and he had truly earned the accolades that came his way. His work reflected well on him, on the CSA, and on this country. He was a leader, a mentor, and yes, a hero to all who looked at his wide-ranging accomplishments and wished they could be more like him. And perhaps most important of all, he was looked up to by thousands of youngsters who hoped to someday become astronauts themselves.

But Williams, who as *Maclean's* magazine points out, "has a CV that makes mere mortals feel distinctly inadequate,"[1] is modest, quiet, and to anyone who has interviewed or written about him, straightforward, and accommodating. Like us all, he is a product of his past, and in his case, that past embodies what was once called the two solitudes of this nation. Dave Williams is a Western Canadian by birth, and a Quebecer by upbringing. He was born in Saskatoon, and educated in Montreal. Then he did postgraduate studies in both Canada and the United States, practiced medicine in several cities, and by the end of his CSA tenure became the first Canadian to have lived and worked in both space and the ocean. He is also a husband, a father, a pilot, and a university professor. In fact, he was working in this latter role when he first learned that he had been picked as an astronaut. While the story of his being chosen has been told before, it bears repeating because it is so reflective of the man himself.

The members of the second class of Canada's astronauts were informed of their selection in the spring of 1992, on the afternoon of Saturday, June 6. That day, Doctor Williams was at the Sunnybrook Health Science Centre in Toronto, where he was about to speak to a crowd of nurses about emergency medicine. Despite the fact that the day was a pleasant one outdoors, the business at hand was inside, and those that were there wanted to be there. No one in the audience that day would have known, nor could they have guessed, that they were about to witness something that was highly unusual, and as far as their lecturer was concerned, one of the most significant moments of his life. In fact, they would leave the auditorium an hour later, none the wiser.

When Doctor Williams' pager went off, no one in the room gave much thought to the fact that he stepped into a nearby hallway and answered a phone. "After a brief, non-committal call, he hung up, and without commenting on the interruption, began his lecture. No one else in the room knew he'd just been informed that he'd been selected … as one of four new Canadian astronauts."[2]

Nor did he show it, largely because the Space Agency swore to secrecy those chosen that day. "I was trying not to look excited," he told a reporter later, "because I was not supposed to let anyone know this was happening. But inside I was jumping up and down."[3]

That phone call, like the ones received by Williams' new astronaut colleagues, came as a welcome relief after long, tension-filled weeks of waiting. Right after his lecture, and as soon as he could reasonably extricate himself from the nurses around him, he phoned his wife and told her the wonderful news. Then he did his best to internalize the significance of the phone message, and how everything it embodied would impact on him and on his family. But one thing he was sure of: Cathy Fraser, his wife, would be supportive in whatever transpired. She was an Air Canada pilot and had not attained success in her profession without being dedicated, determined, and professional. Now she would be as thrilled as he by the good news of his selection, even though neither of them could quickly grasp what it might mean, or where it might ultimately lead. Ironically, it was she who had suggested he apply for the astronaut job when the first postings appeared in the papers.

Dave Williams was still a toddler when his family left Saskatchewan. That has never taken away his affection for the Canadian West, nor from the admiration that the people of Saskatchewan feel about "one of their own" having reached the pinnacle of success in arguably the most glamorous profession there is. The prairie media outlets have always carried glowing reports of his space launches, and of his work, both on the shuttles and out of them.

When he was a boy, Williams was interested in, and attracted to, the larger-than-life individuals from the United States who were going into space and exploring all the new unknowns that it embodied. However, like so many from this country who wanted to become astronauts, Williams found that his desire to explore the same frontier was not an option for Canadians. As we know, all that changed with Marc Garneau in 1984, but until then, Williams had to put his hopes on hold.

"When I was growing up in Canada, I remember watching the original Mercury astronauts on television. And of course, in those days, having a TV was just something incredible in itself, and a little black and white image and you'd watch the amazing flights of the original astronauts, small little capsule, short duration missions, but seeing images of the earth for the first time from space. That was what really captured the desire in me to become an explorer."[4]

At that point, because he could not go into space, the teenaged Williams turned to another unknown that was equally enticing. He resolved to explore the world under the surface of the sea. Again, what he saw on television was a factor in this decision.

Around that time TV documentaries involving the nautical exploits of French oceanographer diver Jacques Cousteau were widely broadcast. Dave Williams had seen them all, and had been entranced by the possibilities they presented. In tandem with the Cousteau exploits, the work of a well-known Canadian also played a role.

Doctor Joseph MacInnis was an Ontario boy from Barrie, who, after he graduated from medical school at the University of Toronto, went on to an illustrious career involving many undersea exploits. MacInnis was the first human being to dive under the ice at the North Pole; he was the man who discovered the largely-intact remains of the ship Breadalbane that sank in northern waters in 1853; he was a consultant to the Titanic discovery team; and he participated in a great number of undersea expeditions in this country and elsewhere. In short, his work and his fame greatly influenced Williams, who decided to do something about it. He learned to scuba dive at the age of thirteen. Then, in due course, he became a doctor and, ultimately, an astronaut.

Canadian astronaut Dave Williams participates in underwater simulation of extravehicular activity (EVA) in the Neutral Buoyancy Laboratory, near the Johnson Space Center at Houston, Texas.

But first, he embarked on what to most people would have been several careers, all of which earned him satisfaction, praise, and collegial respect. He did postgraduate work on advanced invertebrate physiology at the University of Washington, in Seattle. Later, he became interested in a wide variety of aspects of emergency medicine, everything from the evaluation of the retention of cardiopulmonary resuscitation skills, to studies of patient survival after heart attacks, to the efficacy of tetanus immunization in the elderly. His work in critical care had ever-expanding roles: from serving as an emergency physician, to the training of ambulance attendants and paramedics. For varying periods he was an assistant professor of surgery at the University of Toronto, medical director of the Westmount Urgent Care Clinic, director of emergency services at Sunnybrook, and, after becoming an astronaut, manager of the Missions and Space Medicine Group in the Canadian Astronaut Program. No wonder his presence and skills would be welcomed on board a space shuttle during an emergency.

Doctor Williams first flew on *Columbia* on STS-90. That flight is often referred to as the Neurolab mission, because of all the medically-related studies that were done on it. Dave Williams played a vital role. Shortly after that successful mission, he was made director of the Space and Life Sciences Directorate at the Johnson Space Center. In assuming that role, he became the first non-American to hold a senior management position within the NASA organization. Surely this was a long way, in every way, from his birth in Saskatoon, Saskatchewan on May 16, 1954.

On the morning of April 17, 1998, Dave Williams, the man who at one time thought he would become a marine biologist because no Canadians were flying, roared aloft on *Columbia*, and less than ten minutes from the pad at Canaveral, found himself weightless in space. The launch itself, although delayed for a day because of a mechanical glitch, was flawless when it came. On board were seven souls, four of whom were doctors. Dave Williams was one of three Mission Specialists. Kay Hire, also a Mission Specialist, was the sole woman on board. Back in the province where he was born, the Regina *Leader-Post* told its readers that "Canadian astronaut Dave Williams rode a bright-orange plume of flaming rocket fuel into the heavens, during a spectacular liftoff of the space shuttle

Columbia."[5] The paper also mentioned that "among his musical selections for the trip [was] the Canadian rock classic, 'Oh What a Feeling.'"

As soon as the shuttle reached orbit, every member of the crew set about performing the many tasks at hand. Even though there was much to do, and much accomplished, Dave Williams did not cite his work when he was asked about his favourite parts of space flight. His response reflected the personality of the man: "For me, I think, there are two favourite parts. One is simply being there and enjoying the experience, and floating, looking out the window at the sun setting and listening to 'What a Wonderful World,' for instance, or 'Imagine' by John Lennon. It's an absolutely incredible experience to float around looking out at the earth going by underneath you and just take it all in, the spectacular beauty of our planet. The other element, which is as exciting for me, is working together as a team."[6]

The team on STS-90 was extraordinary. There were so many professional scientists involved that it came as no surprise to NASA that the mission was as successful as it was unique. In the laboratory in the cargo bay of *Columbia* was an incredible menagerie of critters, all of which were playing a specific role in the advancement of medicine and science. Among them were rats, mice — many of which were pregnant — snails, fish, and hundreds of crickets. As well, the human subjects on board, were prepared to, and did, involve themselves in several experiments. The spaceship was as unique as any laboratory anywhere, but in this case it was a lab that circled the world every ninety minutes.

The liftoff occurred at 2:19 on a Friday afternoon, and it was seen by the usual thousands and thousands of enthusiastic observers. Among them were many who began their weekend early and didn't care if the boss condoned or cared. The launch was the most exciting thing that happened that day in Florida, and being present was far more memorable than watching the same thing on television at home.

Among the throng that day were members of the Williams family. Cathy Fraser brought their son and daughter, and she was completely enthralled by what she saw. Their son knew what was going on, but his little sister was too young to understand. Nevertheless, they all watched as *Columbia* soared from sight; until it was a tiny speck in the brilliant blue of the sky over the Atlantic. Then, the shuttle was gone. Later on, Fraser described the moment for a reporter.

"After I started breathing again, I just felt overwhelming joy," she said. "I wasn't particularly scared; I was just really excited." And in a somewhat more reflective mood she explained what the launch and lead up to it had been for her: "This whole thing has been a very positive experience and very enjoyable. I'm a firm believer that this is really a safe adventure and I feel really exhilarated."[7]

As she was speaking these words, her husband and the rest of the *Columbia* crew were already halfway around the world, stowing their launch gear, and learning to adapt in an environment that only a select few have ever known. The first moments in space were unforgettable for them all, so much so that later they would be at a loss for words in trying to describe the wonder of the world they had entered. In researching this book, our astronauts often told me how difficult it was for them to adequately explain to the earthbound what it was like to be in space; how amazing it was, how thrilling, and yes, at first, how unpleasant to have to cope with feelings of nausea. Having experienced dives, rolls, and tight turns in a fighter jet, I felt I had a passing understanding of what I was being told. However, a fighter jet soon lands; spaceships go on for days and days. A significant number of astronauts experience space sickness, particularly at the beginning of missions, but they know they can cope with the discomfort, and that within a couple of days the problem generally disappears. They also know NASA has little interest in making the subject a high priority in press releases.

In all, some twenty-six individual life science experiments were done during the STS-90 mission. Every member of the crew participated, while the test animals and insects that were brought along played the roles required. Many of the animals, such as the rats and mice, did not complete the trip. They gave their lives for science in some of "the most complex animal dissections ever done on orbit, removing the brains of rats so scientists could see how the nervous system changes in weightlessness."[8]

By the end of the flight, most of the rodents were killed and their body parts preserved for later study in university labs. Dave Williams and his American colleague, Doctor Jay Buckey, performed the majority of the dissections in which the animals were "beheaded one at a time with a tiny guillotine or given an anesthetic overdose in a sealed laboratory chamber with glove-like openings for astronauts' hands."[9] There were dozens of snails and fish that were killed as well, but many of the crickets on board would be dissected after the return of the shuttle. They had been brought on the flight in order to allow researchers back on Earth to attempt to determine if being in space had affected them in any way.

Because the mission was intended to advance neuroscience research, the effects of microgravity on the brain and the nervous system were principal focuses. That was why the brains of the rats and mice were used, and it was why the humans on board became test cases for other long-planned and highly anticipated studies. In fact, eleven of the twenty-six experiments involved shuttle crew, although for perhaps obvious reasons, the doctors on board were the ones most directly involved. In fact, Dave Williams was not only the crew medical officer — the first on a shuttle — he was, along with his colleagues, a test subject as well.

One of the experiments involved the sleep that the astronauts were able to get during the flight, and how weightlessness affected their rest. Various sensors were attached to the bodies of those being studied, and while the sensors might have been bothersome and not conducive to ensuring sound sleep, they were accepted because they were necessary. Other studies involved spatial awareness, hand-eye coordination, and the like. Attention was given to the differentiation of up from down, which is often initially disquieting to humans in space. In the early part of the period in orbit, Dave Williams mentioned the spatial factor in an interview he gave during a downlink telecast to York University students in Toronto. In answer to a query about the disorienting effects of weightlessness, he explained: "I felt like I was standing on my head on earth where all the fluids shift from your lower extremities and your head feels very stuffy and congested."[10] Then he added that getting around in space was "very similar to being under water." And as we know, a great deal of astronaut training is done under water — in the massive pool built for that purpose near the Johnson Space Center in Houston.

The effects of microgravity can cause astronauts to confuse a floor with a ceiling, or vice versa, and "these sensory balance barometers gone awry may also explain why the elderly suffer from falls."[11] It was that kind of very common earthly effect that Dave Williams and his colleagues hoped to learn more about. That was one of the reasons why he regarded the Neurolab mission as being so important. Right after trips into space, astronauts sometimes have trouble walking, until they get accustomed to gravity again. They suffer bone and muscle loss in orbit, and the longer the mission, the more profound that can be. In a similar vein, bone and muscle wasting are often problems for the elderly. It was hoped that a mission such as STS-90, and the experiments done on it, would supply medical data for treatments on Earth.

In addition, the operations done on the rats and mice were intended to assist astronauts in acquiring proper techniques for emergency treatment of colleagues, in case of injury or accident while in orbit. Such things as a cut that was bad enough to require stitches would require specialized attention where gravity was not present. In fact, Dave Williams alluded to such a problem during a panel discussion in Ottawa several months before the flight. He said it was important to learn how to treat patients in unusual environments, and then added: "That's what's exciting to me — to be a doctor in orbit, understanding how to treat humans who get ill. How do we deal with simple things like cuts and nosebleeds, or appendicitis? We're going to go on after the space station back to the moon, and after the moon, to Mars. To be able to support missions like that, we have to understand how to treat conditions in space."[12]

In the years since Dave Williams made his first trip into space, much more research has been done on most of the matters that the Neurolab mission highlighted. By the time he got the chance to go on his second flight, the lab itself had been retired; the great multinational space station was nearing completion, and the scientists at NASA and elsewhere were more optimistic than ever that problems arising on long-term space deployment were solvable. But all that was still in the future. As far as STS-90 was concerned, this ninetieth shuttle mission was essentially over.

The noon hour skies over KSC were clear as *Columbia* landed on Runway 33, after orbiting the Earth 256 times. Commander Richard Searfoss glided his craft to a gentle touchdown, 1,694 feet past the threshold of the concrete strip, and then rolled 9,949 feet farther along, until the spaceship came to a complete stop. Then it was quickly surrounded by service trucks of all kinds — including medical transport vehicles. Inside there were stretchers that were used to carry Dave Williams and four of his colleagues off the shuttle.

But not because they were ill; the five crewmembers were carried off *Columbia* so that their weightless state could be preserved as long as possible, for scientific study. They were given immediate and extensive examinations, the results of which were intended to assist medical personnel on the ground in determining the effects of gravity — and the absence of it — on the nervous system. The sooner the five could be studied, the more reliable the examinations and the results of them would be. In all, the group would spend six hours being checked over before the initial round of tests were completed. Then the fliers would have real reunions with family members; not just the greetings permitted as the medical examinations were about to begin.

Because the day of the landing — Sunday, May 3 — was a special one for Dave Williams and his wife, both were particularly conscious of it. It was their twelfth wedding anniversary, but out of necessity private celebrations had to wait until the medical examiners had completed their work. Cathy Fraser had a fleeting chance to see Williams though, and reporters asked her about him. She told them that he was in good spirits, but he said that "his arms felt very heavy and he found he couldn't move his head too quickly because it felt like the room was spinning. He wanted to know about the kids, and then wished me a happy anniversary."[13]

And then he was gone, to be pricked, prodded, measured, observed, and examined, all in the name of science. An hour later, he was telling those around him how good a hamburger would taste.

When Dave Williams retired from active astronaut status, he was feted, praised, and his exploits in space were widely acknowledged. His work under the sea was mentioned less often, and that is unfortunate. During the time that he was involved with marine examination and exploration he gave that work as much undivided attention as he did to the furthering of medical research in the space shuttle lab. In the underwater facilities where he worked, he was equally conscious of how medicine might be practiced there. The constraints might not have been as pronounced as they were in space, but they were still highly restrictive.

The United States has an underwater research facility in Key Largo, Florida. A program there that is affiliated with NASA is called NEEMO. Astronauts there are involved in missions of varying lengths, the briefest being a week. They spend the time in Aquarius, an undersea research habitat where they learn to live and work underwater. Williams explains its purpose: "We use that as an analog to help astronauts train and prepare to go on board the International Space Station, but we also use it as a testbed to develop technologies before we send them into space. One of the great technologies we can access there is remote medical care."[14] In today's world, perhaps more than ever before, medical care can be an urgent need anywhere — from the peaks of our highest mountains, to the depths of our deepest mines. As Dave Williams postulates: "How can you deliver health care in a totally isolated environment?" His answer is both practical and succinct, and it involves "being able to use high-speed tele-communications technology, whether it's satellite technology or whether it's fiber-optic cabling, to enable physicians to communicate with either other physicians, other health care providers or, in many cases laypersons to deliver health care in remote, isolated environments."[15]

If anyone, anywhere, is cognizant of what all this means for the future, it is Dave Williams. He spent much time in space, as we know, but his eighteen day mission as Commander of NEEMO was perhaps just as significant. Indeed, his contribution to science and medicine on Earth, above it, and underwater has been second to none.

12

Building a Cruise Ship in the Middle of the Ocean in a Storm

To call Julie Payette a human dynamo is neither exaggerating nor patronizing. She is just that. She is energetic, resourceful, and highly intelligent. Still in her mid-forties, she has done more things, succeeded in more ventures, and contributed more to society than most people twice her age. And she never stops, nor does she intend to. As this is being written she is training for her second space flight, and barring unforeseen circumstances will board *Endeavour* for a mission to the International Space Station early in the summer of 2009. She participated in assembling that facility back in 1999, the first Canadian to help do so. While in space, on STS-96, she circled the globe 153 times in nine days, nineteen hours, and thirteen minutes. Her next flight will be considerably longer.

Payette is married, a mother of two, and a truly accomplished human being. In addition to being an astronaut, she is a pianist, a jet pilot, an engineer, a skydiver, and has sung with the Montreal Symphony, the Piacere Vocale in Basel, Switzerland, and the Tafelmusik Baroque Orchestra in Toronto. She is also a member of the Ordre national du Quebec, has honorary doctorates from fifteen universities, and speaks six languages, one of which is Russian. In her spare time she plays racquet sports and scuba dives. She is also a delightful person to interview.

Julie Payette was born in Montreal, the daughter of a bookkeeper mother and an engineer father. Like so many youngsters around the globe, she watched the *Apollo 11* landing on the moon, and immediately decided she wanted to be an astronaut when she grew up. But, unlike so many who had the same dream, hers came true; though not without a lot of work, worry, dedication, and drive.

And not without a home atmosphere that was conducive to her aspirations. Payette has always been thankful to her parents who she says, "were always behind their children, loving them, supporting them, and encouraging them to go beyond their potential."[1]

Julie Payette was a Mission Specialist on STS-96 in 1999. She will fly again on STS-127, scheduled for 2009.

Courtesy Canadian Space Agency

"I just tried to make a success of my life," she told me in the first of two interviews in did with her at Cape Canaveral. "I am not afraid of work, and I've always wanted to prove myself in whatever I did."

By any criterion, objective or otherwise, it is obvious that Payette has indeed proven herself. As a schoolgirl she figured out quite early that in order to have any chance of becoming an astronaut she needed a good background in science or math. For that reason, she studied both subjects extensively, and applied herself diligently. "I am an organized person," she explains, "but sometimes there does not seem to be enough time to work everything in. That was the way it was when I was growing up, and it has not changed since."[2]

The Payette work ethic was instrumental in her application of what are often called "extra curricular" objectives. A natural athlete, she found that sports were an outlet she loved as a young girl. In the summer she was on the tennis court; in winter the ski hill.

At a time when we are accustomed to hearing that major sports figures had to leave home at an early age to get the best coaching and competition to further their careers, we sometimes forget that such circumstances often apply to others as well. Julie Payette was one such person. She left the family home when she was just sixteen, and did not even remain in this country. Instead, after she was awarded an international academic scholarship, she won admission to a prestigious school overseas and studied there for two years. Her destination was Wales, at the United World College of the Atlantic. She was one of only six Canadians admitted the same year.

All the Canadians, as well as the other students there, were bright achievers, and hard workers. Julie thrived in the environment, and it put her in good stead a couple of years later when she began taking courses that led to her first degree in mechanical engineering at McGill. After that there was a Master's in computer engineering at the University of Toronto, another year overseas working in Switzerland, and finally back to Canada to design speech-recognition software for Bell-Northern in Montreal. Through it all, her dreams never died, and when she noticed newspaper advertisements for astronauts, Payette applied.

"When I was young, I was always interested in planes and flying," she told me. "From about age ten I used to tell my friends I was going to be an astronaut someday. They used to humour me, and I don't think they ever believed it. So when I saw that ad, I applied right away."[3] She says this with a slight French accent, her eyes flashing, and an ear-to ear grin across her face. She gestures as she talks and, outwardly at least, does not seem to take herself too seriously. Nevertheless, Payette is all business when it comes to her chosen profession.

Months passed between the posting of the "astronauts wanted" newspaper notice and the selection of the finalists. The weeding out period seemed to take forever, but finally fifty potential candidates were picked, and for reasons never fully explained, were named and introduced to the media. To Payette, though, the tactic was rather transparent in its intent. "I knew the press conference was to see how we would behave," she said later. "There was no point in presenting the 50 semi-finalists otherwise."[4] This explanation is likely correct. In their professional role, all astronauts have to deal with reporters, and having them do so, even before the final selection, gave the Space Agency committee a chance to evaluate candidates in what, for some, was a high pressure situation.

The fifty hopefuls were then subjected to even more intensive testing, evaluation, and stress-inducing examination. Some were eliminated for physical reasons, such as sight or hearing inadequacies. Others were dropped because it was felt that they were overly aggressive or, conversely, just too laid-back. But for Payette and two men whom she would later know well, Chris Hadfield and Dave Williams, all the testing led to a successful end. Another finalist, Robert Stewart, was selected, but as we have indicated earlier, chose not to pursue a career as an astronaut.

On the day that the final selections were made, Julie Payette was in Hamilton. She and the others on the short list were informed that they would be getting a call from Arline Marchand, the human resources head at the Canadian Space Agency. Marchand would make the calls beginning at 1 p.m. on Saturday, June 6, 1992. Payette got her call at 1:15, and was told that she had just become Canada's second female astronaut. Looking back on the day of the selection, Payette admitted to being nervous as she waited for the phone to ring. "It was a long hour before that," she said. And as she recalled the months leading up to the day, added that "it was the uncertainty that was hard to live with. This meant a big change — I was either branching into this other career objective I had, or I was completely changing my life and becoming an astronaut. I needed to know, and as soon as I knew, everything was very calm."[5]

At last, the slight, vivacious, and determined Julie Payette was headed for Houston for training and the beginning of the new and most exciting job of her life.

The training was relentless, taxing, and purposeful. Along with the others, she spent hours and hours on the reduced-gravity aircraft. She took deep sea diving lessons, got her commercial pilot's licence, and learned to fly CT-114 Tutor jets at the Canadian Air Force Base in Moose Jaw, Saskatchewan. Today, she has well over 1,200 hours of flight time. She completed her initial astronaut training in April 1998, and by the time STS-96 was set to launch a year later, would be a Mission Specialist on the flight. The months, and then weeks, leading up to departure were a blur of activity, final preparation, and long days of training. Not long before the mission was to commence, Payette found herself flying a small aircraft near the launch site. She preserved the moment in an email to her younger sister Maude: "Things just hit home big time when I flew a Dash-4 over the pad this evening at Kennedy Space Center. There it was — *Discovery*, all white and shining under the late afternoon sun. I just couldn't believe my eyes. Twenty-four days before launch date and here stood the very orbiter I will ride to space or to oblivion. But ride the mighty rocket, ride I will."[6]

And she did. "My launch was in the early morning," she told me, "so we had to be up and out to the pad in the middle of the night. You saw the shuttle floodlit at night, so you know how beautiful it is." I had, but I was not about to ride it. Having seen this magnificent piece of machinery did prompt me to ask if she would like her children to see such a sight, to someday ride the rocket themselves. "Oui," she answered quickly, "but the decision will be theirs. And I will be supportive in anything they do." And you know that she will.

Shortly before launch day, NASA staged a press conference at the Johnson Space Center to introduce members of the shuttle crew to the media. This was a chance for journalists to question the astronauts before they flew up to Kennedy and into quarantine before their flight. Because the mission was an important one, and because *Discovery* would be docking at the International Space Station (ISS), media interest was quite high. The assembled throng of reporters crowded into the auditorium and jostled for position as they awaited the crew. When the astronauts, dressed identically in red, short-sleeved sweaters with the NASA symbol imprinted in white on the left chest, had taken their places on the dais at the front of the room, the questions began.

As each reporter identified him or herself, and indicated the organization for which they worked, it soon became obvious that the centre of attention was Julie Payette. She, according to one journalist, was "the biggest star." Then the same man added: "It helps that she is attractive, vibrant, and does not shy from the spotlight." This was why, "she was the focus of attention as she and her fellow crewmembers fielded questions at their last public appearance before going aloft. Eighteen of the 22 questions were directed to Payette — as the five Americans and one Russian who will fly with her sat smiling and mostly silent."[7] By the time the session ended, the young astronaut from Montreal had the press eating out of her hand, and they continued to do so in their reporting throughout the entire flight.

STS-96 was an important mission. Packed into the shuttle payload bay were two cranes that would be transported and attached to the ISS. Along with the cranes were tools, handrails, even foot restraints necessary for future assembly of the ISS. Several items of cargo were included, among them, storage batteries that would replace others that were already at the ISS. Two Americans, Tammy Jernigan (who had flown with Steve MacLean) and Dan Barry, would be conducting spacewalks, while Julie Payette's task would be to choreograph these walks from the shuttle's flight deck. When asked by a reporter about the building of the ISS, over time the biggest construction project ever contemplated for space, Payette replied that it would be similar to "assembling a whole cruise ship in the middle of an ocean during a storm."[8] In all, forty-five shuttle flights would be needed before the 108 metre-long structure would be complete.

The actual launch of STS-97 was supposed to be earlier than it was, but the fuel tanks of *Discovery* were damaged in a hailstorm that swept across central Florida a week before blastoff — a portent of what would befall other shuttles. NASA immediately moved the launch to a later date. This time, on the morning of May 27, all was in order.

Long before the scheduled hour, hundreds of spectators sought out the best places to watch the spectacle. The early morning air was unusually humid and uncomfortable, but that did not deter the crowds. Area roadways were lined with cars, invited guests were wending their way towards Kennedy Space Center viewing positions, and local boaters used their knowledge of the adjacent waterways to move as close as they could to the pad. In almost every case, whatever the size of the boat; it was full. Several entrepreneurs hooked up with strangers from out of state or even out of country, to take them to the "best viewing sites" — for a fee. In researching this book, I talked to an individual from Atlanta who drove to Titusville, a town near KSC, because she and her husband often wished they could see a launch. They did not know, or care, who the astronauts were; the launch was the attraction. At Titusville, they found a man who would take them in his boat, and guaranteed that they would not be disappointed. He transported them then, and to five launches since. None were disappointments, and over time they've all become good friends.

A short time before blastoff, one local boater managed to circumvent Coastguard patrols and got inside a well publicized security zone. Luckily, his sailboat was spotted in time and the shuttle countdown proceeded normally. The boat owner was ordered to move elsewhere promptly; he had been right in the area were the shuttle's rocket boosters would fall.

For each shuttle mission, families and friends of astronauts are grateful for invitations they get to watch the spaceship depart. This was certainly the case when Julie Payette flew. Her parents and other family members, including her sister Maude, were there. Maude was sought out by reporters for her and her parents' reactions to the whole thing. Nearby were two nuns who taught Payette at College Regina Assumpta in Montreal. Sisters Annette Bellavance and Jacqueline Villeneuve had been more than just her teachers; they were also Julie's dear friends. Doing her best to keep a low profile that morning was Aline Chrétien, another friend, and the wife of Canada's then prime minister. Everyone waited with anticipation and keen interest in what was about to happen.

Then, as one reporter wrote,

> [The] parents of Canadian astronaut Julie Payette wept as they watched their daughter thunder into the early morning sky in space shuttle *Discovery*, riding a plume of flame and smoke into orbit.
>
> Spectators burst into applause, and a large Canadian contingent cheered as the shuttle lifted slowly off the ground in what NASA called a 'picture perfect' launch, climbing skyward before the backdrop of a just-risen sun.[9]

A short while later, Julie's sister Maude told anyone who asked that she thought the blastoff was "magnifique." And it was; for those who were there, and for those who were no closer than their television screen. In Montreal that morning, fifteen members of the religious order, the Congregation de Notre Dame, got up early and watched the spectacle before they went to mass. Afterwards, they told newsman Mark Abley that they were thrilled by the launch, and they told him what their astronaut graduate was like when she was a student at their school.

Sister Yolande Perrault said she had warm memories of many former students, from United Nations war crimes prosecutor Louise Arbour to Julie Payette. Sister Yolande, who taught Julie French in grade eight, remembered her well. "She was very attentive and she worked tremendously hard. I remember one day there was some time between classes, and the other girls all went outside to relax. But Julie stayed in the classroom to study." The former teacher also recalled Payette's "dedication, her level of energy, and the fact that she was interested in everything."[10]

Payette took that drive and dedication with her into space. There was a lot of work to be done once *Discovery* attained orbit, and this young Canadian woman was determined to contribute in every way that she could.

The main purpose of the flight was to dock with, and deliver supplies to, the International Space Station; *Discovery* astronauts anticipated the work ahead. Mission Commander Kent Rominger was sure that he could do the docking, and had trained for months for the task. At the time, there were no inhabitants on the space station. The supplies would be for the first crew who would go to live there. Once *Discovery* arrived at its destination, two spacewalks would be required before the transfer of goods began. These walks were inherently dangerous. The astronauts would be tethered at all times to *Discovery*, but in case one of them "got loose and floated away, he'd ditch the station and chase after the astronaut,"[11] said Commander Rominger. Fortunately, he did not need to do so. Mission Specialist Julie Payette supervised the spacewalk and operated the Canadarm. The hookup with the ISS proceeded as planned. During the visit, the spacewalkers attached two cranes to the side of the station; they would be needed during subsequent missions.

In the period after the docking, *Discovery* crewmembers entered the space station, which at the time seemed both austere and noisy. The noise came from on-board fans. In order to lessen the racket, foam insulation was hauled through the connecting tunnel from the shuttle and permanently installed in the station. Because the place was still not occupied, its two rooms lacked the warmth they would have once long-term residents arrived. In all, some four tons of supplies were transferred from *Discovery*. These included battery packs to replace ones that had ceased to function. A broken radio was repaired as well.

Because all shuttle orbits are not the same, astronauts see different areas of the Earth on different missions. In that sense, Julie Payette was particularly thrilled by the trajectory on STS-96. The shuttle passed over Central and Eastern Canada, and even though the glimpses she got were all too fleeting, she was able to see Montreal from space, as well as the Magdalen Islands, the Gulf of St. Lawrence, and all of Nova Scotia. Being able to do so was deeply satisfying, and it increased her understanding of one of the real joys of being an astronaut. Later, she would often allude to what she saw on that flight and would mention how beautiful the Earth was as it hung in the black and endless universe that surrounded it.

During the flight of *Discovery*, as well as before and after it, Payette was lauded by the media, particularly the Quebec outlets. Because her background was in that province, her every move, word, and opinion was reported, and as might be expected she later became the star attraction at every gathering she attended there. However, the Quebec media were not alone in singing her praises. News organizations in the rest of Canada did as well, and several reported what Commander Rominger had to

Courtesy John Melady

Julie Payette pictured at the Kennedy Space Center during the summer of 2007.

say about those he flew with, and about Canada's youngest and newest astronaut. "The crew has been fantastic," the veteran flier told an audience at Ottawa's National Museum of Technology in a broadcast from space. "Julie Payette in particular, has been a wonderful addition."[12] One of those who heard the remark was Prime Minister Jean Chrétien, who in turn praised Payette, and wished her well.

In due course, the three and a half days that *Discovery* crewmembers spent inside the space station came to a conclusion. The long and difficult closing of the six hatches between the two spacecraft was accomplished, and the shuttle undocked, moved away from the ISS, and set course for home. For Payette, her trip of a lifetime was ending all too soon.

The return flight to Cape Canaveral was described by the press as rather routine. For those inside the shuttle, however, the trip never lacked for excitement. Later, the crew gave various descriptions of the voyage, but none of these was more enthusiastic that Payette's.

"It was a fantastic adventure from beginning to end," she told assembled reporters who gathered for a post-flight press conference in the media compound at KSC. "The re-entry is spectacular, especially when you get into the atmosphere and see the plasma on fire and the shuttle is surrounded by that. We were a big glider coming down and de-accelerating from 25 mach. It was an extraordinarily smooth landing, to the point that nobody was sure that we had touched down."[13]

As the session concluded, someone asked Payette if she would like to go into space again. Her answer was non-equivocal: "I will continue to work with NASA and hope I get a second mission," she responded without hesitation.

Now her wish is about to come true. During a media briefing at the Canadian Space Agency in Montreal, on February 11, 2008, the word was out. Julie Payette will be a Mission Specialist on STS-127, and will be returning to space, and to the International Space Station in 2009. She has truly earned the chance to fly again.

13

Like Christmas Eve When You Are Seven

M arc Garneau's first space shuttle flight came about more quickly than either he or anyone else involved in the Canadian astronaut program had expected. But he had to wait almost twelve years for his second mission. In the interim, *Challenger* happened, a new group of Canadian astronauts was picked, and Bondar, MacLean, and Hadfield all flew. In the time between his missions to space, Garneau qualified as a Mission Specialist, served as a capsule communicator (CAPCOM) — he was the first non-American to do so — and had been named Deputy Director of our astronaut program. He had also remarried in 1992, five years after the death of his first wife Jacqueline. During those years, he "faced all the complexities of being a single father of teenaged twins while juggling a demanding job that required a great deal of travel."[1] He succeeded admirably, and when the time came for him to embark on his second mission, in May 1996, he was better prepared, much more familiar with NASA and its code of operation, and eminently more aware of what to expect and how he could contribute, both in training, and on board the shuttle after it left the pad. In short, he was a veteran flier, and it showed.

He was more self confident, more prepared to question poorly thought out procedure, more adept at problem solving, and more realistic in his expectations for himself and for those around him. As one of his former colleagues told me: "Marc was an ideal astronaut, and he was so totally focused, it was unbelievable. He worked damned hard always, and just presumed everyone else should be doing the same." These traits had always been part of his persona, whether as a student at the Royal Military College in Kingston during the late sixties, or completing his Doctorate in electrical engineering at the Imperial College of Science and Technology in London. As a naval officer, and then as an astronaut, the man was a leader in the truest sense of the word. He has a keen mind, the ability to synthesize complex and disparate arguments quickly, and his organizational skills are second to none. On a personal level, he responded to my emails with succinctness and dispatch — sometimes inside of an hour. The fact that he was chosen for a second mission surprised no one.

The flight itself was known as STS-77, on the spaceship *Endeavour*. This was the shuttle that was named following a countrywide contest in the United States, open to all elementary and secondary school students. The idea was to give the shuttle the same name as a ship that had been involved in a research role at some time in the past or, more likely, a vessel that had been famous historically. U.S. President George Bush Sr. announced the winning name in May 1988. "*Endeavour* was named after a ship chartered to traverse the South Pacific in 1768 and captained by 18th-century British explorer James Cook, an experienced seaman, navigator and amateur astronomer. He commanded a crew of 93 men, including 11 scientists and artists."[2] The work done by Cook is well-known, particularly his charting of the waters off Australia and New Zealand, and the naming of many plant and animal species that had been hitherto unknown.

Because Cook's ship was British, the spelling of the name for the shuttle retained the British variant — *Endeavour* — with the "u" intact. Some Americans told me they were annoyed by this, and as late as the STS-118 mission in the summer of 2007, were insisting the shuttle should have an "American" name. In fact, just before the August 8 launch, a huge fifty-foot banner "Go Endeavor" was plastered across the high, razor wire no-go barrier at the base of pad 39B at Kennedy, where the *Endeavour* stood. A day or so later, the *Orlando Sentinel* ran a photo of the sign with the spaceship in the background. The paper captioned the illustration with the note: "OK, the sign is missing a 'u,' but the spirit is there."[3]

There was a crew of six on board *Endeavour* when Garneau flew for his second time; all were male. The shuttle was under the command of American Colonel John Casper, a former Vietnam War combat pilot, and a veteran of three previous missions. With him on the flight deck was Pilot Curtis Brown, who was on his third shuttle trip. There was only one rookie, a Mission Specialist named Andrew Thomas from Adelaide, Australia. Another Mission Specialist, Mario Runco Jr., had been a state trooper in New Jersey before he became an astronaut. The other crewman was Daniel Bursch, who had flown before, and like Garneau, had a navy background.

All were highly trained, and since the *Challenger* disaster were required to have a particular skill that was deemed less important earlier on. After *Challenger*, an emergency escape system had been devised for shuttle crewmembers, allowing them to parachute to safety in case of a near-Earth accident. Whether the plan would work was highly debatable, and was never used, but the parachute training became de rigueur nonetheless. In addition to the safety factor, advocates of the move insisted that it was a good way to determine who could handle stress and who could not. It was a matter of record that Canadian astronauts had always advocated the training, even before it became mandatory. In fact, our first team members "managed to persuade CSA officials to allow skydiving as well, arguing that since their job requires them to spend quite a bit of time flying in high performance aircraft, they should know what to do if they had to bail out."[4] One of those advocating for the move was Marc Garneau. His reasoning was that "you discover something about yourself in the moment you actually jump and fall out of the sky. It teaches you about your ability to work in potentially stressful situations and how to react quickly and correctly."[5] His assessment was sound, and no astronaut interviewed for this book felt otherwise. In fact, they all loved the thrill, and the adrenaline rush. Unsurprisingly, they all loved the excitement of the launches as well.

Despite having flown before, Garneau was just as enthusiastic about this flight as he had been about his first. "It's like the excitement of Christmas Eve when I was seven years old," he told a reporter shortly before the mission began.[6] That certainly seemed to be true; those who worked with him during the final preparations for launch noted that he seemed happier than they had ever seen him. The long wait between flights was finally over. The years of work and preparation were about to pay off, and the ever-present fear of a scrubbed launch was about to ease. "It is a challenge to have your launch date slip continuously," he once remarked. Now though, there was no slippage — the *Endeavour* would go on schedule.

The flight was an important one in several respects, but particularly to NASA for reasons that were not fully exploratory. In fact, the organization was then, as it often seemed to be, short of the funds it needed to continue the programs they felt to be necessary for the future. Government cutbacks were always feared, and became an ever-present factor in how many missions were flown, and what the purpose of each would be. In earlier years, the U.S.–Soviet competition, and the race for the moon, meant a well-funded NASA. Now, thirty years on, the potholes in the roads and the other needs of their districts meant that members of Congress were more inclined to look after their own, rather than doling out ever-decreasing resources for a space program that, to some, was a bottomless pit. Even though some Congress members fully supported what NASA was doing, there were others who looked upon the launching of rockets as a colossal waste of money. That was why, whenever possible, the public relations types at NASA went to great lengths to stress the commercial aspects of the missions, and to point out to the taxpaying public just how many benefits would come from what was being done.

In fact, "in an era of government spending cutback, space program supporters, more than ever, are emphasizing the practical benefits of costly ventures beyond the Earth's atmosphere. Of course, the National Aeronautics and Space Administration — a media darling of the sixties heyday — has always put a premium on public relations, and it has continued its hard-sell approach as its glow and its funding have faded. NASA has billed its latest mission as the opening of 'the commercial space frontier.'"[7]

That was why, on STS-77, in addition to the science that was being done, purely commercial objectives were also a factor. The most blatant example was a project involving the Coca-Cola Company and its attempts to spread the dogma about their product. In an earlier, highly-publicized advertising scheme, Pepsi-Cola had sponsored a pitch for their drink: Russian cosmonauts extolled its goodness while they were on board the space station Mir. Now it was Coke's turn to make their pitch, but in a true spirit of orbital one-upmanship, they were also attempting to perfect a kind of soft drink dispenser that would operate in microgravity. Those most familiar with space operations were not surprised to learn that NASA even had a pretentious, and rather silly, name for the machine: a "Fluid Generic Bioprocessing Apparatus." However, once the thing was in space, it wouldn't work.

As might be expected, there was a measure of criticism about such blatant advertising, so the reasoning for it had to be defended. That rather thankless task fell to a man named Ed Garbis, the then director of space processing for NASA. He explained the rationale: "Coca-Cola is a premier U.S. capitalistic company which is seeking to improve their competitive advantage. Part of our charter is to extend the interest of commercial firms in space."[8] Whether the explanation satisfied many is debatable, particularly because NASA was providing over three million dollars to underwrite the scheme.

Another objective of the mission was the deployment of a Spartan experimental satellite with a Mylar antenna attached to it. The antenna would be folded up initially, but would expand to the size of a tennis court once it and the satellite were removed from the shuttle cargo bay. If the antenna worked, the theory was that it would help reduce the cost of solid structure antennae on the still-to-be-built International Space Station. At the end of the experiment, the Spartan would be retrieved and the antenna released to burn up as it entered the Earth's atmosphere.

Also on this mission, a module called Spacehab was positioned in the cargo bay, where the astronauts would be able to enter and leave it through the tunnel connected to the crew quarters. Spacehab was where several other life science experiments would be conducted. Having the lab with them meant that the astronauts had a much larger area in which to work and live.

Among the less publicized studies conducted on the mission were two with obvious medical ramifications. The first was an investigation of whether an insulin-like growth factor would help reduce bone loss in space. The second was an attempt to produce a crystalline form of a new type of insulin. There were also embryonic starfish, mussels, and tiny sea urchins in an on-board aquarium. These miniscule life forms would be studied to try and gain more information on birth defects and osteoporosis in humans. In essence, despite most of the publicity, criticism, and justification centering around the temperamental performance of the soft drink machine, there were important studies being done on the mission. No wonder Canadian-directed work on computer chips received less attention.

Right up to launch time, those around Marc Garneau noticed how enthusiastic he was for what was about to transpire. A reporter asked him to compare the mission with the one he went on earlier.

"The first time I didn't know what lay ahead of me," he explained. "This time I do. Because I know what it's like, because it's so wonderful, I'm really excited about it."[9] He went on to mention how lucky he had been on his first flight, because the shuttle had passed over Canada several times. Now, however, the inclination of the mission differed, and the path followed around the Earth would be more to the south; south of the Great Lakes. But even so, he added, "when I look out the window, I will see some of Canada, but I will not see the west, it's too far north. I will however, be able to see Toronto in the distance."[10]

The launch of *Endeavour* for the STS-77 mission took place at 6:30 a.m. and, according to those who were there, was essentially flawless. Included among the usual thousands of spectators were family members of those on board, senior NASA officials, politicians, and representatives of the local, national, and international media. Marc Garneau's wife, Pamela, had her thirty-sixth birthday the day before the launch, and she made particular note of the occasion, explaining that the safe launch was indeed a special present she would never forget.

Nor would the other witnesses; among them, thirty-six students from Marc Garneau Collegiate in Toronto. Their school, named after Canada's first astronaut some ten years earlier, had sent the students to the Cape on a field trip that was truly unique. Their enthusiasm showed, and as the shuttle blasted off they could hardly believe what they were seeing. "I was bawling my eyes out," admitted nineteen-year-old senior, Mike Davies. "All my life, I have dreamed of becoming an astronaut, and this was just a coming together of everything I've always dreamed of. This was a mission with a Canadian astronaut ...who fulfilled his dream the way I want to."[11]

Ashley Waltman, a vice principal at Garneau, was one of those who accompanied the students to Florida. He was just as impressed by the spectacle: "The power, the brightness, the light, the rolling thunder that washed over us — that's what I can't get over. We were about five miles away, and none of us can get over the way the sound came at us, building, building until your internal organs were vibrating. It was very moving and very powerful."

Because Marc Garneau had visited their school and some students had met him, they had an increased interest in what they were seeing. One was a grade ten student who said that because he had talked to Garneau, he could not help but think of the astronaut as the shuttle roared into the distant sky. "It's something to see a shuttle launch," enthused grade ten student San Gennidakis, "but it's something else entirely to think that you actually know someone who's aboard.

"You think about what he's accomplished in his life, and it makes you look at the shuttle launch in a whole new perspective. Some of us are beginning to realize 'Hey! Maybe that's something I want to do,' and it encourages you to start learning and focusing your abilities."[12]

Even before Gennidakis made those comments, *Endeavour* was already in space and the crewmembers were busy going about the things that needed to be done. First off, according to the bulletin issued by NASA, was the "work activating the Spacelab module systems in the cargo bay laboratory and the ship's robot arm."[13] It was the robot arm, or to Garneau, the Canadarm, that he would operate during the flight. However, the arm was not used that Sunday, the first day in space, because by the time the crew had familiarized themselves with their new environment, had stowed their launch equipment, and had completed all the necessary primary tasks, their work day had come to a close. After all, they had been up since the middle of the night in Florida, and despite their excitement, badly needed rest. Their first eight-hour sleep period began at 4:30 in the afternoon, Central Time — the time at Mission Control in Houston.

As had been planned, the deployment of the Spartan was one of the first operations on the first full day on orbit. The large, box-like, gold-coated satellite and the folded antenna were lifted from the cargo bay by American Mission Specialist Mario Runco. Several tests were done and then it was released. The operation went well, and once released the giant inflatable antenna expanded to its full size — over thirty feet in diameter. The following day, Marc Garneau used the Canadarm, retrieved the Spartan, and gently lowered it back into the cargo bay of the shuttle. Operating the Canadarm was gratifying for him, and afterwards he said it was "almost an extension of your body. It is a real pleasure to use."[14]

As the mission continued, so did the experiments in Spacehab. Some of them involved Canadian-designed elements, and even projects initiated by elementary and high school groups in various locales. One, proposed by youngsters at College Park Elementary in Saskatoon, looked at the way liquids diffused in microgravity. Another, suggested by students from Toronto, studied the way a human could throw objects in space. Needless to say, the originators of the schemes were highly interested in them.

And then there was the Coke machine....

Despite various attempts to make the thing work, all of the initial efforts failed. The theory behind it dealt with the mixing of fluid and gas in weightlessness, but the practicality was questionable. Finally, attempts to get the machine to operate began to pay off. Midway through their fifth day in space, the two navy men on the shuttle, Dan Bursch and Marc Garneau, took matters into their own hands. After lots of trouble-shooting, delay, and annoyance, they saw some improvement. There was much tinkering, colourful comments from both, and lots of joshing from their colleagues. Then, whether because of, or in spite of, all the good-natured teasing and less-than insightful suggestions, the two achieved success. Down in Houston, Mission Control announced the good news as part of the daily status report. In the bulletin for the day, Houston even elaborated: "The dispenser was then tested by the crew and currently [it] is working well and filling drink containers."[15] The operative word there was "currently," and no further mention of the machine was made for the balance of the trip.

On the final day of the mission, Prime Minister Jean Chrétien called Marc Garneau on *Endeavour*. The two chatted for a few minutes, and Mr. Chrétien congratulated the astronaut on a successful mission. The prime minister also expressed his pride in all the experiments that were carried out. South Australian Premier Dean Brown offered similar sentiments to Adelaide astronaut Andrew Thomas. Then, their conversations with the politicians complete, both fliers resumed their work in wrapping up the science projects and began securing the crew quarters for the return to Earth.

At the time, there were four possible landing opportunities: two different times in Florida and two others at Edwards Air Force Base in California. The preferred site was on the East Coast, where an early morning return to KSC was acceptable, prior to the arrival of showers that were predicted to move over the Cape as the day progressed.

In due course, STS-77 came to an end. *Endeavour*'s breaking rockets were fired at 5:09 Central Time, "to enable the shuttle to drop out of orbit for its hour-long slide back to earth. *Endeavour* streaked across the Pacific, the San Francisco Bay Area, the Rocky Mountains and the Gulf Coast before crossing over into Florida to align itself with KSC's Shuttle Landing Facility."[16] The touchdown was without incident.

While it had not been announced at the time, Marc Garneau would soon begin preparations for a return to space. Again, it would be the shuttle *Endeavour* that would take him there. It would be his final mission.

14

Waving the Canadian Flag at the World

On the last day of November 2000, Marc Garneau returned to space. He spent 240 hours there, operated the Canadarm, and directed three spacewalks at the International Space Station. Then he returned to Earth, participated in all the post-mission medical checks, debriefings, and NASA duties, then retired as an astronaut. He soon became president of the Canadian Space Agency, but his flying days were done.

Garneau's last mission was on *Endeavour*. The designation this time was STS-97, with five souls on board: the Commander, Pilot, and three Mission Specialists. All had been in space before, and at fifty-one, Garneau was the oldest member of the crew. Four and a half years had passed since his previous mission.

The main purpose of STS-97 was to deliver and install the first U.S. solar rays at the Space Station. Prior to their being put in place, the facility was greatly underpowered and parts of it were essentially unusable. The solar rays themselves were folded and compact in the shuttle cargo bay prior to liftoff, but by the time they were removed in space and attached to the station, would be large, elegant, and integral to the fledgling development being constructed.

The flight to take them to the heavens was the 101st shuttle mission. It left Pad 39-B in Florida at 10:06 p.m., and the trail of fire that hurtled *Endeavour* from Earth could be seen for miles in every direction. Departure was on time, and entry into space was about nine minutes later. By that time, responsibility for the flight was through Mission Control in Houston, and would be for the balance of the journey. Henceforth, all bulletins about progress would be datelined Houston, and would be on Central Time. At the time of the launch, the Space Station was over the Indian Ocean, and almost eight thousand miles ahead of *Endeavour*. In due course, they would come together in a rendezvous in the sky. This meeting was perhaps most anticipated by three long-term residents, one American and two Russian, who were already on the station and had been for a month. For them, a few new faces would be more than welcome.

Because everyone on the shuttle had done this before, they all knew what to expect once on orbit. As always, the hours before blastoff were devoted to the last minute preparations for the flight, but each astronaut also did what he could to put himself at ease in anticipation of the always long, exciting, and tiring launch day. Shortly before the mission began, Brian Tobin, Canada's then industry minister, phoned Marc Garneau to wish him all the best and a good flight. Afterwards, the minister told the press about the astronaut's demeanor: "He sounded very confident, very serene, very proud to be waving the Canadian flag at the world."[1]. Garneau was not only the sole Canadian on board this flight; no mission has ever carried more than one Canadian. There have, however, been several instances where the CAPCOM in Houston has been Canadian while a fellow countryman flew.

As the *Endeavour* streaked through space towards its link-up with the International Space Station, industry observers noted that this was the first time in almost three decades that two crewed space vehicles were under NASA's control at the same time. Those who were old enough will recall the days when a space capsule was landing on the moon, while a second vehicle waited for the lunar travellers to return. The most vivid instance of this, of course, occurred when Neil Armstrong and Buzz

Aldrin descended in Eagle onto the lunar surface, as Michael Collins, in the command module *Columbia*, circled the moon and waited for them.

This time the moon was not the objective, and the rendezvous of the spaceships would be nowhere near it. In this case, the planned link-up was 230 miles above northeastern Kazakhstan. The docking would occur first; the unfurling of the huge, electricity-producing solar wings would take place second. The wings themselves were manifestly utilitarian, and on Earth they would have been capable of producing enough power for at least thirty homes. In space they would prove to be a vital component in the long-range operation of the space station, which, at the time, was projected to cost upwards of $60 billion when fully complete. By then it would also be one of the brighter objects visible in the heavens at night. These solar wings were also the largest ever carried into space.

News accounts of the same month in 2000 mentioned the flawless operation of the shuttles, and the regularity of trips into space. The missions themselves had approached routine in the wake of the increased safety precautions after the *Challenger* accident. Of course, this was before the equally disastrous loss of *Columbia*, but that was still in the future. Nevertheless, at the time that Marc Garneau was winging his way towards the space station, more and more Americans and Russians were being trained for missions. In view of this situation, there was increased anticipation that another recruitment campaign would soon be necessary in Canada as well.

At the time, Michel Vachon was general manager of the Canadian Space Agency, and reporters asked for his reaction to the speculation. He did not attempt to downplay it. In fact, he concurred with the widespread rumours and drew attention to the obvious increase in the number of missions that were being flown, and the belief that many more lay ahead. To that end, he said a new hiring program would indeed be needed.

"The big question is to know when to do it," he said. "When we know how many flights we'll have over the next ten years, we'll know how many astronauts have to be recruited."[2]

What Vachon did not know then, nor could he have known, was that a host of obstacles, monetary and otherwise, would hinder human exploration during the ensuing months and years. In fact, the recruitment campaign that seemed so close at hand at the time he spoke, did not become a reality for almost a decade. It finally opened in May 2008. However, the positive spin concerning such a project was not out of place when it was made, the timing was just off. But at the time of STS-97 all was well.

As the *Endeavour* sped through space, periodic firings of the shuttle's jet thrusters lined the ship up for the anticipated link-up with the ISS, in two days time. In the meantime, the cargo bay doors were opened, the Canadarm was checked, and the space vision system was tested to ensure that all would be ready to go as soon as they were needed. Meanwhile, the three men who waited on the space station prepared their domain for the arrival of their visitors. Their anticipation was great.

Docking day was December 2, and the actual coupling of the two spaceships occurred at 2 p.m., Houston time. Just two hours later, Mission Control announced: "Canadian Space Agency astronaut Marc Garneau maneuvered *Endeavour's* Canadian-built robotic arm and grappled the 45-foot-long, 17.5 ton P6 solar array truss structure at 4:17 p.m., lifting it out of its berthing latches in the shuttle's cargo bay, where it will remain overnight attached to the arm to properly warm its components."[3] During the docking, while Garneau worked the arm, the three spacemen inside the station strained to see what was happening on their doorstep. As they were doing this, the two Mission Specialists in the shuttle with Garneau were making their preparations for the necessary spacewalks to properly install the solar wings.

Fortunately, the wings were unveiled and installed successfully. The operation was done over two days and involved two spacewalks. On a succeeding day, fine-tuning of the operation necessitated a third trek outside of the shuttle. While those feats were being performed, back on Earth the media were posting accounts of the progress. There was joy in Houston as the maneuvering continued. In the words of an Associated Press story, picked up by other outlets, the pride of accomplishment was obvious:

> Space shuttle *Endeavour's* astronauts attached the world's largest, most powerful set of solar panels to [the] International Space Station yesterday, then watched with delight and relief as the first glittering wing unfolded.

The blue-gold panels, made of silicon cells and thin Kapton layers, were folded like an accordion for lift-off. [The] wings, covering 2,000 square metres, will be the largest structure yet deployed in space. Each wing — 11.6 metres wide and covering 32,800 solar cells — has power storing batteries and radiators at the base. The combined wingspan — 73 metres — exceeds that of a Boeing 777. NASA expects the big U.S. wings to generate 65 kilowatts at peak power.[4]

The two Mission Specialists whose outside work made the installation possible were pleased that their efforts had gone well. One of the two, Joe Tanner, a year younger than Marc Garneau, had done two separate spacewalks on an earlier mission. With the addition of three on that mission, and two more six years later, he would eventually spend over forty-six hours in Extravehicular Activity (EVA). His colleague, Lima, Peru-born Carlos Noriega, eventually accumulated over nineteen hours.

While the two men were out of the shuttle they wore helmet cameras, devices that are commonplace today, but were being used for the first time on that mission. Typical of any new or revolutionary device, the camera Tanner was using stopped working before he completed his walk. The one on Noriega's helmet functioned reasonably well. Technicians in Houston were able to follow the wing installation progress through it. From inside the shuttle, Marc Garneau was in radio contact with Noriega, and advised him how to move his head — and the camera — so that the images could be picked up by Mission Control. In a sense, Garneau functioned somewhat like a film director as he choreographed the operation. When the spacewalk was complete, the second important part of the mission began.

The shuttle and the space station flew for six days in locked position. The same link-up made it possible for the men on both spaceships to visit each other. These meetings, with the expected backslapping, laughter, and camaraderie, were not just social affairs. Various supplies and equipment were delivered, including a pair of vice grips that one of the men on the station had requested. The same man, American Bill Shepherd, was also overjoyed when another of his wishes — some fresh coffee — was handed to him. A new laptop was also delivered, as were headsets for two-way video conferencing, a new hard drive for a Russian computer, and large containers of water. In a seemingly unbalanced return gesture, containers of refuse from the station were moved to the shuttle for a return to Earth. The long-anticipated get-together was especially appreciated by the space station inhabitants. When the time came to do so, they seemed almost reluctant to say goodbye.

The space station–shuttle link-up also made possible several broadcasts from space; on the big American networks, cable news, Discovery Channel, and Canadian media outlets. As with most projects involving both television and space, the interviews were scheduled well in advance, and timed to the exact minute. For example, three separate broadcasts were done, one after the other, beginning at precisely 2:31 p.m. CST on December 4. The purpose of these exchanges was to allow the astronauts to tell viewers on Earth about the focus of the mission, and its progress at the time. Shortcomings, if any, were not mentioned. The spin that NASA wanted was, not surprisingly, to be positive. After all, it was the taxpaying viewers on Earth who were footing the bill. Reports of a mission in trouble might enhance television ratings, but such an eventuality would not ensure popular support for the costly and dangerous work being performed by the men who were circling the Earth at seventeen thousand miles an hour.

As expected, Marc Garneau participated in the media exchanges, as did all of his colleagues. One of these joint sessions involved reporters at the Kennedy Space Center, the Johnson Space Center, and the Canadian Space Agency near Montreal. Despite attempts by some of the reporters to elicit complaints about STS-97, none were forthcoming. The answers from space were polished, professional, and positive.

This was in contrast to a Garneau interview that took place shortly before the mission even began. In answering a reporter's questions then, he was completely honest, blunt, and surprisingly forthcoming. In the exchange, he talked about what it was like to be a rookie astronaut seventeen years earlier. He mentioned that when he arrived in Houston for the first time, he was looked upon as an outsider who somehow barged into the closed environment that was the U.S. Space Program at that time. He felt that he was being watched, and in an unwelcome way.

"I felt eyes burning through my back when I walked down the hall," he explained. "It was very much an American program." He said he had the feeling that non-Americans "weren't quite part of [it]." He recalled that he had to be told where to stow his sleeping bag on orbit, how to warm his meals, and even how to use the zero-gravity toilet on the shuttle. Nevertheless, he persevered, but was always conscious of his actions, particularly during that first flight. "There was a lot of pressure," he added. "I wanted to acquit myself right."[5]

As history would show, he did. In due course, Garneau became a Mission Specialist, participated fully in NASA operations, and played an integral role in the two missions that followed. He would never again get the impression that he was performing at a level below that of his American colleagues. In essence, he had arrived, was at ease, and was fully accepted in the roles he was assigned to carry out.

At the time of Garneau's first mission, the Canadarm was already a vital component of the shuttle. Garneau saw the arm, of course, was awed by its effectiveness, and noted the controls that were used to operate it. Yet at the same time, he could not use those controls, even though they, and the arm that they moved, were Canadian, built in Canada, and were paid for, in part, with his own tax dollars. In fact, because he was untrained to operate the arm at the time, he was forbidden to even touch the controls. That part, he found, was "especially humiliating."

Finally, on STS-97, he was the Mission Specialist whose prime responsibility was the operation of this great Canadian invention. But, "my hand was shaking a little bit at the start," he admitted.[6]

Among the broadcasts from space involving Marc Garneau, was a link to the Canada Science and Technology Museum in Ottawa. Gathered there were Girl Guides, navy cadets, and students from two Ontario schools. These youngsters watched, fascinated, as "their" astronaut demonstrated what it was like to be in a micro-gravity environment, and answered questions about a variety of subjects. One query dealt with space travel, and the possibilities for its future. Garneau assured his listeners that each of the seventy-five young people in his earthly audience would see unprecedented, exciting, and rewarding space travel in the years to come. They might not participate themselves, but some in their generation would.

"Within my lifetime, and certainly within your lifetime, we're going to send humans to Mars," he declared.[7] Many in the audience nodded in agreement, and some no doubt envisioned themselves being part of such exciting adventures in the years to come. Meanwhile, the adventurers on STS-97 started to prepare to return to Earth.

The hatches between *Endeavour* and the International Space Station closed on Saturday afternoon, December 9, in preparation for the undocking. This was done over the northern reaches of the Persian Gulf. The three men remaining aloft shook hands with their visitors, said their goodbyes, and wished the departing fliers a successful and safe journey. Then the shuttle moved away from the station, and using the thrusters again, *Endeavour* flew once around the ISS and set course for the return. At the point of separation, the two spacecraft were approximately 240 miles above Earth. The trip home — through space, into the Earth's atmosphere, and subsequently back to Florida — would take two days, with touchdown occurring in the December darkness of early Monday evening.

Even though no official announcement had been made at the time, in his comments during one of the news conferences from space, Marc Garneau said that he expected that this mission would be his last one. "You can't get too greedy in life," he pointed out. "I've fulfilled just about all of the things that I would have ever dreamt — even many that I didn't even dream I could ever do.

"So I'm very, very happy to have been able to fly on three occasions and especially to finish off with an opportunity to visit the [space] station." Then he added, in what seemed to many a somewhat bittersweet remark: "Even though this is a fantastic experience, I think it's going to be my last flight. At some point in life, you have to make changes."[8]

Soon afterwards, the space shuttle landed, and Marc Garneau, Canada's first man in space, stepped off *Endeavour* and into history.

15

The Most Memorable Thing an Astronaut Can Do

Aside from stepping onto the surface of the moon, spacewalking is arguably the most memorable thing that an astronaut can do. Chris Hadfield, Steve MacLean, and Dave Williams were the first Canadians to have performed this remarkable and dangerous feat. The three Mission Specialists were all involved in spacewalks during the construction of the International Space Station. While carrying out their allocated duties there, all of them exhibited more resourcefulness, stamina, and courage than most of us can imagine. They used the skills they possessed, overcame the obstacles they faced, and made significant contributions to the building of the massive structure in the sky. Despite the lonely, hostile, and ultimately unforgiving environment in which they worked, the three never failed in their resolve; never retreated, and in the face of tremendous odds, never left any task undone. Station construction benefited from their expertise.

Chris Hadfield led the way. In 1995, he flew on STS-74. Then six years later, on April 19, 2001, he returned to space on STS-100. Hadfield's earlier mission was on *Atlantis*. This time, the spaceship was *Endeavour*, and the trip would be three days longer than his first. On *Atlantis*, he had orbited the Earth 129 times; this time he would complete 186 rotations. There were five crewmembers on that first trip; on the second, seven. In both instances, Hadfield was the lead Mission Specialist.

STS-100 blasted off from Kennedy at 2:41 p.m., Florida time, and entered orbit less than nine minutes later. At the same time, the International Space Station was high above the Indian Ocean, and the three crewmembers there were informed of the successful shuttle departure. In fact, about twenty minutes after the launch, they were even able to see a video uplink of the event, provided for them from Mission Control in Houston. The ISS crew had been in place for just over a month, and when the newest shuttle crew docked with them, the visitors from it would be their first.

The STS-100 mission was a special one for all Canadians. Not only was there a fellow countryman involved, but the main purpose of the flight had an important Canadian component. Carefully secured in the big payload bay of the shuttle was the fifty-seven-foot-long robotic invention — Canadarm2. This Canadarm was being transported to, and installed on, the Space Station. Chris Hadfield and American Mission Specialist Scott Parazynski would do the physical work of putting the device in place. With upgrades and the necessary maintenance it would operate from there for the life of the station. From the date of its installation, the arm would become, and remain, one of the most useful and critical components of the entire edifice. No wonder this mission was highly anticipated by NASA, by the Canadian Space Agency, by the thousands of men and women who built the arm, and by those who take an interest in space and in the advances being made there. In essence, the installation of Canadarm2 was a highly complex and utilitarian step into the future.

Not only was the manufacture of the arm a long and intricate construction project; its installation also necessitated thousands of hours of training by the astronauts who were involved in placing it where it was needed. A great deal of that training took place in what most people would regard as a huge swimming pool.

What NASA calls the Sonny Carter Training Facility/Neutral Buoyancy Laboratory was opened near the Johnson Space Center on May 19, 1997. This structure, often referred to simply as "the pool," was named after astronaut Sonny Carter, who developed many of the spacewalking techniques that ultimately became widely used. The popular and inventive Carter was in training for his second shuttle mission when he was killed in a commercial plane crash on April 5, 1991.

The massive pool, a hundred feet wide and twice as long, is forty feet deep. In NASA-speak, it provides "controlled neutral buoyancy operations to simulate the zero-g or weightless condition which is experienced by the spacecraft and crew during space flight [and] is an essential tool for the design, testing and development of the space station and future NASA programs."[1] In other words, astronauts train in the pool, and by working underwater they learn how to anticipate their roles in space. They "wear their full spacesuits along with weights to keep them underwater. Specially trained personnel in scuba gear are always nearby in case something goes wrong. Inside the pool are full scale replicas of the shuttle payload bay and space station modules."[2]

As the shuttle sped towards its linkup with the space station, the *Endeavour* crew went through the usual stowage of the equipment that was needed during the launch, adjusted to being weightless, and, when possible, took quick glimpses out of the eleven windows of the ship. The payload bay doors were opened en route, and would remain so until it was time to come home. In the bay itself, Canadarm2 was ready for deployment, as was an Italian Space Agency cargo module called Raffaello, which contained several tons of equipment for the space station.

As the journey continued, the *Endeavour* crewmembers prepared themselves in every way possible for the tasks ahead. To a person, they felt that they were as ready as they ever would be. When asked, Chris Hadfield's reply pretty much summarized the positive aspects of the moment. "I wasn't worried at all," he recalled, "I had trained for four and a half years, and I knew that what I was expected to do would be a tremendous personal, technical, and physical challenge. But no, I wasn't worried about it."[3] He went on to mention that he and his colleagues were looking forward to reaching the station and fulfilling the expectations of the mission.

Courtesy NASA

Astronaut Chris Hadfield stands on the end of the Canadian-built Canadarm
to work with the Canadarm2 during its installation on the International Space
Station, April 2001.

About the same time back home, a newspaper reporter was asking Chris Hadfield's brother, Dave, about the qualities he had recognized in his astronaut brother. The answer that the writer got reflected the positive elements that Chris would demonstrate in carrying out the duties that awaited him. Obviously, Dave Hadfield was entirely confident in his brother's abilities, in his quiet professionalism, and in his utter hesitancy to boast about his achievements. "The qualities that make you successful are not those that attract attention," Dave Hadfield explained. "You need self-discipline, perseverance, and all around intelligence as well as a healthy body. None of these are standouts by themselves."[4]

However, taken together, they represented what writer Tom Wolfe determined was "the right stuff." Indeed, NASA's reliance on Chris Hadfield in this mission was a vote of confidence in his abilities, and in what he could do.

On the second day of the mission, the flight crew on the shuttle carried out the various rendezvous manoeuvres necessary for a successful linkup to the space station. While that was being done, the spacesuits to be worn by Hadfield and Parazynski while they were outside the orbiter were checked and re-checked, as was the Canadarm, which would be critical for the removal of its namesake and Raffaello from the cargo hold. By 5:00 p.m. on the same day, the *Endeavour* was some 1,400 miles below and behind the space station, and was gradually closing the gap by 170 miles every time it orbited the Earth. Finally, just before nine o'clock on the morning of Saturday, April 21, the docking occurred. At the time, the two spacecraft were southeast of New Zealand, at an altitude of 243 miles over the southern Pacific Ocean. Then, on Sunday, April 22, 2001, the first spacewalk by a Canadian began. Chris Hadfield floated out of the airlock on *Endeavour* at 6:45 a.m., CST. Much later, he told me what it was like.

"Because I was the lead spacewalker, I was carrying the burden of responsibility for the entire endeavour. I was completely technically prepared for this walk. I knew right down to the hand motion what I was planning to do for the six or seven hours we were going to be outside. We had back up plans for back up plans."[5]

But he was not ready for what came next. "I was not prepared for the overwhelming visual assault that you are subject to when you are outside.

You have the entire universe in front of you. It is stupendously visually powerful and stimulating. It is the most beautiful thing you have seen in your whole life. You want to be able to tell people to be quiet and look at this thing. And it is like that all the time! It is mind-numbing to see, and then to try to tear yourself away from this raw, unprecedented beauty all around you and focus on the job you have been training for is physically difficult, and an insult to what is happening all around you. And you have to take some time and soak this thing up because it is so magnificent. It is a rare and tremendous human experience, and you can't ignore it. I felt I had to take some time and look at it and try to absorb as much as I could so I could make a slight attempt at trying to tell others about it.

"I had to force myself to go back to work. All the while you are outside, the clock is ticking and you are very aware of it. Since everything else but the visual is contained inside your suit, the only real link you have that you are on orbit is the visual."

But the necessary duties did get done, and Hadfield was extremely cognizant of his role in upholding the Canadian end, in representing his country well in everything he did. This aspect of the mission was further reinforced that morning, shortly before he and Parazynski left the shuttle confines. Steve MacLean sent a message of congratulations to Hadfield, assured him of his support, and wished him all the best. Even Mission Control Houston got into the act by transmitting a rousing version of "O Canada," performed by Roger Doucet, the beloved police officer who used to sing the national anthem at Montreal Canadiens hockey games. No wonder Hadfield was determined to do his best. The pressure was unrelenting.

Once the magnificent shock of being at one with the universe was somewhat absorbed, the two spacewalkers turned their attention to matters at hand. Their first task was to install an ultra high frequency antenna (UHF) on what was called the Destiny module of the space station. The twenty-eight-foot-long Destiny was an American laboratory that had been affixed to the ISS a couple of months earlier. The deployment of the antenna took two of the seven hours in the first spacewalk. But then, just as the two men were getting more attuned to their surroundings and becoming more confident in their abilities outside the shuttle, Hadfield found himself embroiled in a situation he could never have predicted. In fact, his problem was unprecedented, unexpected, and serious — but not even mentioned in NASA's official status report that covered the day.

Chris Hadfield suddenly went blind!

What happened was mentioned, rather casually, by Hadfield himself, in answer to my question regarding methods of rescue in case an astronaut became incapacitated during a spacewalk. When I asked the question, I had no idea how close to home it actually was.

"You can become incapacitated for a lot of reasons," he explained, "so we practice how to deal with whatever the problem is: right from someone being sick and throwing up inside their helmet, to having their vision messed up, to having a hole in their suit, and maybe dying. We practice how we are going to get that person back inside. We practice in a virtual reality lab and go through what we have to do to get the person. You have to get them tethered to you; then you have to move them around in more or less the same way you would move a five hundred pound payload. If they happen to be on the Canadarm, you might be able to fly the arm around to the airlock, so the rescue would be quicker. We train our arm operators to be able to do that. And then the other spacewalker has to stuff them into the airlock. We have to get the hatch closed and get them depressurized. I think we could probably do it in about thirty minutes, but that would be about as quick at it would happen. I was blinded, and I could not see for at least half an hour when I was on my first spacewalk."

The remark was unexpected.

"I was working away outside," Chris Hadfield told me, "and suddenly my left eye became irritated. Then it started tearing up and stinging, like when you get raw shampoo in your eye. Your eye shuts and you have to rub it and flush it because you can't see out of it anymore. Well, my eye started doing that. Because I could not figure out what was causing the problem, I tried to work with one eye for a while, and I didn't tell anybody. The trouble is, tears need gravity, and drain because of gravity. That way, your eye cleans itself; it generates tears, which wash the contaminant from your eye, and when there is gravity causing the tears to fall, you are okay.

"But without gravity, you just get a bigger and bigger ball of contaminated tear. Then it gets big enough so that it goes across the bridge of your nose and gets into the other eye. That's what happened to me. Soon both of my eyes were blinded and I couldn't see out of either of them," Hadfield explained. As he did so, he paused in his narrative, as if temporarily reliving the experience. His facial expression told me that describing the situation — even long after the fact — was not particularly pleasant.

"I opened and closed my eyes, again and again, but they wouldn't clear, and the situation actually got worse. Now both were totally clouded with the contaminant. Finally, I called Houston, told them I couldn't see, and that I would have to take a break." Meanwhile, Hadfield was well over two hundred miles above the Earth, isolated from his partner, gripping the shuttle, and travelling through space at ten miles a second. Yet, even with the discomfort and the uncertainty of what was happening, he refused to panic, and was able to talk to Mission Control in the same kind of casual matter of fact manner that airline pilots use when they welcome passengers on a flight.

"But Houston was quite worried because our suit gets rid of carbon dioxide by using a chemical called lithium hydroxide, and we get our oxygen through our backpack, and it flows through a lithium hydroxide canister. But if the chemical breaks through its filter and gets into the suit, one of the first symptoms is eye irritation. It's also bad for your health because it can get into your lungs and do a lot of damage.

"So Houston's concern was that I had a lithium hydroxide leak. Because of that, while I was still blind, they asked me to start venting my suit. So I opened the valve for this, and started dumping my oxygen. I remember thinking: this is a really strange place to be. I can't see what I'm doing; I'm holding onto a spaceship, and I'm pouring my life-giving oxygen out into the vacuum of space. But I became very philosophical about it, hoping it would clear the problem after a while. But it didn't.

"So, after about fifteen minutes of that, Houston said: 'Okay, close the valve.' So I closed the valve. But after twenty-five to thirty minutes, I'd cried enough, and there was enough tearing that it diluted the contaminant. Then the tears evaporated. So, taking the place of gravity, the evaporation took place and I could see again, though things were a bit murky at first. They got better though, and I was able to see for the rest of the spacewalk.

"When I came inside, they took a bunch of swab samples of the dried stuff that was around my eyes, and we found out what it was. We had covered the inside of the visor with a surfactant to keep it from fogging up, and the surfactant had a few chemicals in it. I had a water leak from my drink bag, and it had picked up a big bubble of this stuff and that got into my eye. It was nothing life-threatening, but it was mission-threatening. Since then, we have changed the way we clean our visors."[6]

While all of this was happening in space, the media on Earth were generally not aware of just how serious Chris Hadfield's problem was, although they noted it in passing. For the most part, the stories they ran dealt with the size of Canadarm2, the fact that it was essentially a "high tech crane," that it weighed more than a ton and a half, and that it was made of aluminum, steel, and graphite epoxy. The fact that it was folded up in the cargo bay of the shuttle was mentioned, and that Chris Hadfield had once described it as a big spider with its legs curled up. Once unfolded and attached to the space station, the arm would represent a "giant leap for Canada."

Now that he could see again, Hadfield turned to his partner, and together they began the transfer of the arm to its placement on Destiny. Shuttle Pilot Jeff Ashby used the robotic arm on-board to lift the big metal, U-shaped pallet that held Canadarm2 out of *Endeavour*'s cargo bay. Then the two spacewalkers attached the pallet to Destiny, and hooked up temporary power cables so that the "space crane" could be activated and its arms unfolded. While all this was being done, Chris Hadfield was tethered to Canadarm by foot restraints so that he would not float away into space, and so that his hands were free to perform the operations needed.

From time to time during the procedure, Hadfield mentioned the stark, utterly majestic beauty of the world in which he worked. He had emerged from the shuttle when it was over the Atlantic, just off the coast of Brazil, and his first observations were exuberant, like the delight of a child in thrall of what is before him. "Oh man, what a view!" he exclaimed as he exited the airlock for the first time. "That takes your breath away." Later on, he tried to describe the Australian southern lights for his colleagues inside the shuttle and at Mission Control: "The horizon is lit up with tentacles going up a huge distance into space," he said. "They come almost far enough to be under us." Then at one point, he reflected on the whole experience when he said to Scott Parazynski: "When I was a little kid wanting to grow up to be an astronaut, this is what I wanted to do."[7]

By the end of that first day, he and Parazynski had spent seven hours and ten minutes spacewalking, and during that time had circled the Earth almost five times. More importantly, they had completed the objectives for the day, and each had earned a much-needed rest. However, their excitement made resting difficult.

The work the spacewalkers did was noted elsewhere, of course. On their farm at Milton, Ontario, Roger and Eleanor, Chris Hadfield's parents, watched the unfolding of the drama on NASA TV. They were transfixed in front of the flickering screen, and found themselves listening intently for the always clipped and sometimes garbled transmission of their son's words from space. Both were so proud of him, and at times found it difficult to comprehend the wonder of what was happening so far above the clouds.

Senior astronaut Marc Garneau followed the progress of the walk from his location at the Canadian Space Agency facility, near Montreal. The veteran flier told a reporter that he felt two things as he watched his colleague and friend conduct an activity that he himself had not been fortunate enough to have done.

"The first is, boy, I wish I was doing that," he said. "The second is a lot of pride. Chris is fulfilling a dream, and at the same time helping Canada to do something extremely important for us. We have reached a level of maturity comparable to other space nations."[8]

In Houston, Mission Control was ecstatic with the success of the whole operation. When Hadfield and Parazynski ventured outside the shuttle for the second walk, the status report issued by NASA referred to the fact that *Endeavour's* two spacewalkers worked as space age electricians, completing connections that allowed the new International Space Station robotic arm to operate from a base on the outside of the Destiny science lab.

"Hadfield and Parazynski worked to complete all of the primary goals of the mission, including the connection of the Power and Grapple Fixture circuits for the new arm on Destiny, the removal of an early communications antenna and the transfer of a spare Direct Current Switching Unit from the shuttle's payload bay to an equipment storage rack on the outside of Destiny."[9] In those few words, despite their jargon-like tone, it is obvious that the decision makers in Houston were pleased.

The second spacewalk lasted seven hours and forty minutes, and as far as the two men outside were concerned, went extremely well. By the end of their work day they were able to reflect on what had transpired, take pride in it, and see the results of their efforts. Canadarm2 was in place, was operational, and ready, at long last, to make the further expansion of the space station a reality.

Another historic part of the mission for Canada occurred on Saturday afternoon, April 28, when an astronaut on the space station named Susan Helms, who was part of what NASA called Expedition Two, used Canadarm2 for the purpose for which it was intended. The maneuver was described in a status report from Mission Control issued that day: "A Canadian 'handshake' in space occurred at 4:02 p.m. Central time today, as the Canadian-built space station robotic arm — operated by Expedition Two crew member Susan Helms — transferred its launch cradle over to *Endeavour*'s robotic arm, with Canadian Space Agency astronaut Chris Hadfield at the controls. The exchange of the pallet from station arm to shuttle arm marked the first-ever robotic-to-robotic transfer in space. The successful exchange of the pallet was the last remaining major objective of the mission."[10]

While the shuttle was docked at the ISS, the crews of the two spaceships had a reunion of sorts, and everyone not otherwise occupied pitched in and unloaded the over four tons of supplies that were delivered in the Raffaello module. These included racks of hardware intended for installation on Destiny, running slats for a broken treadmill, radiation shields, mufflers for noisy machinery, science equipment, and even clothes and towels. As well, "the shuttle astronauts delivered mail from home, apples, oranges, carrots, celery and candy."[11] In return, Raffaello would be packed with gear and garbage from the station, all of which would end up in recycling depots or dumps once returned to Earth. None of it was jettisoned into space.

The crew of STS-100 returned to Earth on schedule, but they were unable to land at Cape Canaveral as hoped. Cloud, continuous rain, and high winds swept the Atlantic coastal areas of Florida on May 1, the date of return. As conditions were expected to remain unchanged for the days that followed, Kennedy Space Center's Leroy Cain, the entry flight director, waved off landing opportunities there. Instead, *Endeavour* glided to an 11:11 CDT touchdown at Edwards Air Force Base in California. The mission was over. In all respects, it had been successful, historical, and safe.

16

Storms, Malfunctions, and Risk

Someone who is superstitious might claim that Steve MacLean's second flight was star-crossed. It had to be postponed because of tragedy on an earlier mission, deferred because of technical difficulties, scrubbed for a time after a lightning strike, and delayed in the face of a tropical storm. Then, after being on orbit for ten days, had its return held off because of unidentified debris in space and unfavourable weather on Earth. Yet, through it all, after it all, and — in a sense — because of it all, the mission ended successfully, if not on the date intended. The whole enterprise was a testament to the ingenuity of those who prepared for it, directed it, and flew it. For so many reasons, STS-115 was indeed a memorable mission. Yet it came in the wake of an accident that was as terrible as it was unforgettable.

That accident had happened on February 1, 2003, in the crystal clear heavens above southwestern United States. On that day, the space shuttle *Columbia*, with six Americans and one Israeli on board, was headed home after almost sixteen days in space. The oldest shuttle in the NASA fleet was finishing up its twenty-eighth flight. The mission itself had been an important one, involving research in physical, life, and space sciences, and during its time on orbit almost eighty experiments had been carried out. Even though it had circled the Earth 255 times, the 256th rotation would never be completed. As *Columbia* approached the California coast, heading for Cape Canaveral and home, "the 117-ton shuttle was moving through space eight times faster than a bullet from an assault rifle." Then, "in the next fifteen minutes, the shuttle would shed the bulk of that unimaginable speed … enduring three thousand degree temperatures as atmospheric friction converted forward motion into a hellish blaze of thermal energy. It had taken four million pounds of rocket fuel to boost *Columbia* and its crew into orbital velocity. Now the astronauts were about to slam on the brakes."[1]

It was about that time that various on-board sensors indicated a problem on the left wing of the machine. Mission Control in Houston quickly became aware of the matter, but were essentially powerless to do anything about it. On the shuttle flight deck, the situation was noted, and what had been anticipated as a routine return to Earth was not to be. A short time after a member of the crew had taken video photos of her smiling, waving, and outwardly relaxed fellow fliers, the situation with the port wing accelerated. Then, far below, "casual observers on the ground saw disquieting omens in the dark western sky. The orbiter's light track appeared to brighten and shed debris at about 8:54 a.m. Central Time, (5:54 Pacific Time). Over the next twenty-three seconds witnesses reported four similar events, and then a flash as the spacecraft entered Nevada airspace, flying at Mach 22.5 and at a height of 227,400 miles. More of the same continued as *Columbia* flew over Utah, Arizona, and New Mexico towards Texas."[2]

At Mission Control, Astronaut Capcom Charles Hobaugh had been in regular voice communication with the returning shuttle. For that reason, even as he realized the seriousness of the situation on board, he continued to call Commander Rick Husband in the disintegrating orbiter. Suddenly, as Husband attempted to respond to a query from Earth, his message ended abruptly: "Roger, uh…"[3] That was his final transmission, signalling the loss of the spaceship and its crew.

Because *Columbia* broke up thirty-seven miles above the Earth, and was travelling at almost eighteen times the speed of sound at the time, parts came down over a wide area. Nevertheless, the search for the wreckage began immediately, and a board of inquiry to look into what happened was convened almost as quickly. Later that same day, George W. Bush addressed his nation from the White House.

"My fellow Americans," the President began, "this day has brought terrible news and great sadness to our country. At 9:00 a.m. this morning, Mission Control in Houston lost contact with our space shuttle *Columbia*. A short time later, debris was seen falling from the skies above Texas. The *Columbia* is lost; there are no survivors."

The message was blunt, and it did not attempt to sugarcoat the severity of what had transpired. Hundreds, and soon thousands of police, NASA employees, military personnel, and private citizens began searching for shuttle wreckage. "A two thousand square mile debris field, shaped like a cigar, ranged from Fort Worth, Texas, to Fork Polk, Louisiana deposited as *Columbia* disintegrated. In all, about 25,000 individuals participated, recovering some 84,000 pieces of *Columbia* and combing on foot a 700,000 acre landscape."[4]

On a personal note, during my research for this book, I visited the massive Vehicle Assembly Building (VAB) at the Kennedy Space Center. While I was there, I was both surprised and interested when the NASA official escorting me happened to mention all the things that were found that had once been part of the lost spaceship. I remember listening as the man talked about the recovered materials, and how important they were in the accident investigation. Then, when I asked him where the stuff was now, he looked at me and pointed.

"They're right here up there," he said, indicating the upper levels of the building where we stood. And he was correct, of course. In contrast to the parts and pieces of *Challenger* that had been buried on the KSC property, debris from *Columbia* was retained. Now a great bulk of it is stored on the sixteenth floor of the VAB. It has all been carefully catalogued, bar-coded, and sorted for reasons that may not have even been considered yet. By keeping the stuff engineers, investigators, and scientists hope to learn from it; find out why some parts failed, why others survived, and by

Courtesy John Melady

Before being moved out to a launch pad, all space shuttles are assembled on a mobile launcher platform inside this giant Vehicle Assembly Building at the Kennedy Space Center. The VAB is one of the largest buildings in the world, and can be seen for long distances in every direction.

studying both, help insure that whether it be a shuttle or its successor, any machine going into space will incorporate as many safeguards as possible.

There were elements from *Columbia* that assisted in the efforts to pinpoint the cause of the catastrophe. Most were broken, bent, or burned, but taken together they helped investigators determine, with certainty, what the cause of the crash was: a briefcase-sized chunk of foam that had impacted the port wing of the spacecraft during its launch and had breached the shuttle's Thermal Protection System. Then, in the words of the official report on the matter, "during re-entry, this breach in the Thermal Protection System allowed superheated air to penetrate the leading-edge insulations and progressively melt the aluminum structure of the left wing, resulting in a weakening of the structure until increasing aerodynamic forces caused loss of control, failure of the wing, and breakup of the Orbiter."[5]

Fortunately, despite the loss of *Columbia*, the remains of all crewmembers were located and positively identified, and grieving families were able to claim their own. But even in the midst of such sorrow, a chance discovery of two objects that survived the crash did carry elements of consolation. One was the video made by the astronauts just before the orbiter broke apart. Being able to see crewmembers smiling, waving, and working was comforting for their families. The second was a watch that one of the astronauts was planning to present to his girlfriend at the end of the mission. Amazingly, despite the thousands of square miles of search terrain involved, the watch was found, without its face, and with the hands forever frozen at the exact time of the disaster.

After the loss of *Columbia*, NASA went through a period of turmoil. After a short grieving period, there was much soul-searching, finger pointing, and second-guessing; a description of which is well beyond the scope of this account. Eventually, there were constructive criticisms, design changes, increased safety modifications, and, ultimately, widespread resolve to move on. Initially, the entire shuttle fleet was grounded while the cause of the disaster was being determined and the problems rectified. Finally, "after thirty months of work, NASA launched the Space Shuttle Program's Return to Flight — called STS-114 — using *Discovery*. This flight was the most-photographed Space Shuttle flight ever, with numerous cameras on the ground, ships, and aircraft tracking the vehicle during ascent, and the crew of the International Space Station taking a series of detailed photographs as the Orbiter approached the ISS. In addition, the crew of *Discovery* used cameras in the cockpit and on a long robotic arm to examine almost every inch of the Orbiter."[6]

The increased attention paid off. There were still problems that had to be attended to, but all were resolved to the point that a second "return to flight" occurred. This time, more changes were made — more safety factors addressed — until, finally, the first "regular" mission since *Columbia* took off. It was called STS-115, and Canada's Steve MacLean was a Mission Specialist on it.

Prior to flight, as previously mentioned, another host of problems presented themselves. The first was a technical one: an antenna bolt at the launch pad had to be replaced, in a kind of last minute maneuver that was unexpected, time consuming, and extremely challenging.

Nevertheless, the situation was rectified, but it took so long to do that the launch had to be rescheduled. Because of the *Columbia* tragedy, NASA officials decided to launch the first shuttles that followed during daylight hours only. By doing so, potential problems could be detected and (hopefully) eliminated, providing they showed up in the vast array of pictures taken during the launch sequence. Photography was simply more reliable in daylight.

Finally, just when all systems seemed to be in order, and with the shuttle resting on the pad and ready, the good weather needed for the launch disappeared. A vicious electrical storm swept across central Florida and several thunderstorm cells were detected over the Space Center. Then, as a local newspaper reported, "the largest launch-pad lightning strike in NASA history"[7] occurred on Friday, August 25, 2006. The day before the shuttle was to go.

"A powerful thunderbolt struck the lightning mast on the top of the shuttle's 36-story launch tower," the paper continued. "About 100,000 amps of electricity shot down the three-quarter inch wires that run from the mast to the ground, where sensors measure the voltage output. That's five times the electrical current in an average stroke of lightning. It's enough electricity to simultaneously operate 83,333 television sets." How the paper determined that particular number was not clear. What was clear was that the shuttle would have to be examined carefully prior to any launch.

The afternoon following the strike, NASA officials scheduled a news conference to explain the severity of what had happened. I was among those who crammed into the KSC media briefing room to hear senior shuttle manager Roy LeCain offer his assessment. He talked about the storm, showed two black and white, rather grainy, video images of the lightning strikes, and then explained that the launch date would have to be moved back again so that the shuttle could be checked out.

"We see a couple of indications that makes us want to go look at the ground systems and the different flight systems of the vehicle to make sure we don't have problems before we go fly," he said. "We know just enough to know that we don't know enough. We need to let folks go off and look at their data, and that's what we are going to do. We want to be sure we have no technical problems."[8]

In the end, it was determined that the shuttle was in relatively good shape. However, something much more ominous was not far away. A tropical storm called Ernesto was expected to hit the Cape, and the shuttle on the pad would be exposed to whatever nature might have in store. The disturbance had moved over the Atlantic, gained strength in the Caribbean, and, for a time, was expected to enter the Gulf of Mexico and even threaten New Orleans, the beleaguered Louisiana city that had been devastated the previous year. However, the next best guesses about the track of the storm still placed it in the Gulf, then veering north/northeast, and crossing Florida in the direction of the Cape. Already, twenty inches of rain had fallen in Haiti and the Dominican Republic. The rain and gale force winds were not to be taken lightly. Florida citizens, who had been largely spared hurricanes that summer, feared that they were due. Emergency food, water, flashlight batteries, medicines, and the like were set aside, and many shoreline residents began to place plywood over windows. But what to do with the space shuttle?

Atlantis remained on the pad all the following day, largely because weather forecasting is an inexact science, and the National Hurricane Center in Miami was not sure where Ernesto was going. Then, Governor Jeb Bush put Florida on a state of alert, as the winds of the storm increased in intensity. A hurricane watch was issued for the Florida Keys and tourists were ordered to evacuate that area as quickly as possible. Those pulling travel trailers and recreation vehicles were specifically urged to leave the island chain right away.

Meanwhile, NASA launch managers watched the storm, feared the worst, and elected to move the shuttle out of harm's way. As one newspaper told its readers, "NASA has rolled back the shuttle to the shelter of the KSC's Vehicle Assembly Building four times in the program's history because of hurricanes and tropical storms. The space agency's hurricane rules require that the rollback must be completed before sustained winds reach 46mph."[9] Once *Atlantis* was off the pad and inside the VAB, it would take a total of eight days to bring the orbiter out and do the necessary preparations for another launch. In any event, no shuttle was going to be flying for some time.

Then, the rollback was cancelled before it was complete. A series of ever-changing weather forecasts proved to be something of a godsend for NASA. The expected hurricane was downgraded to a tropical storm; the housing of the shuttle was stopped, and one of the great crawlers that was already transporting it to the VAB returned it to Launch Pad 39B. Then, and for the next several days, the orbiter was checked and checked again. Finally, after even more delays because of faulty censor readings, everything was ready. The fuel tanking was done, the crew was belted in, the countdown was done, and *Atlantis* took flight. When liftoff occurred at 11:15 Florida time, on the morning of September 9, the long-delayed return to the assembly of the International Space Station had begun. The astronauts on *Atlantis* were thrilled, NASA was relieved, and the public turned to other concerns.

When the shuttle left Florida, the space station was 220 miles above the North Atlantic, between Greenland and Iceland. As with other launches,

Courtesy NASA

Prior to departing on the spaceship *Atlantis*, Mission Specialist Steve MacLean adjusts his launch suit in the White Room at Launch Pad 39B. The mission he was about to fly was STS-115; it involved almost twenty days in space, his spacewalk, and orbiting the Earth 187 times.

the crewmembers in the sky were able to watch the liftoff via a video link. On board the station at the time were three men: a Russian, an American, and a German. They anticipated that the shuttle would reach them early on day three of its journey.

This mission — STS-115 — was particularly important because its main purpose, as described by NASA, was to "install the 17.5 ton, bus-sized P3/P4 truss section to the station that included a second set of solar arrays, batteries and associated electronics. The addition will double the station's capability to generate power from sunlight. The girder-like truss is 45 feet long."[10] In other words, the section being installed was big, heavy, and although it was not mentioned at the time, costly. The installation would be highly complicated; so much so that the astronauts doing the work had trained longer for it than had any other crew for any mission up to that time.

All the training would pay off, but as Steve MacLean admitted later, the delays in getting airborne had become worrisome. "The day before the launch, I began to wonder. What am I doing? I have a great family, three kids, and I realized I was taking a risk. Granted, it is calculated, but it is a great risk, but one that I accepted a long time before. But then, the day you go, you are so busy with what you have to do; the risk does not occupy your mind.

"Yet if you are not concerned at all, there is something wrong with you. Let's face it; you are sitting on millions of pounds of thrust. And, if the two rockets don't light within milliseconds of each other, it's a very bad day. When the two solids light; there is risk there. If something goes wrong in the first two minutes, there is nothing the crew can do, so that would be bad. Of course, once you are on orbit, there is the risk of a hit by a micro meteorite, but as space is so huge, that probability is low. Then, the first fifteen seconds of re-entry are dangerous. So, if someone says they are not concerned about these things, that doesn't make much sense."[11]

In a news conference at the Canadian Space Agency offices soon after the end of the mission, MacLean expressed similar sentiments. "I actually was pretty concerned before the flight when we were discussing the lightning hit," he recalled. "But once we had all the data, then it was clear to me that we were probably going to be okay."[12] And they were — then, and during the mission itself.

On the morning of their first full day in space, the six astronauts on *Atlantis* were awakened by a recording of Audrey Hepburn singing "Moon River." Shuttle Commander Jeff Brett's wife, Janet, had made the request, and Houston complied. Presumably, Brett liked what he heard, but there is no record as to how his companions reacted. Nevertheless, the crew day was a busy one, with the bulk of their time being spent examining the shuttle for any hint of damage during liftoff.

This examination was, as one newspaper reminded its readers, "a high priority ever since an errant chunk of insulating foam caused the loss of *Columbia* and its crew."[13] The inspection was a thorough one. Using the Canadarm and the fifty-foot-long orbiter boom sensor system, the shuttle's entire heat shield was examined, including the leading edges of the wings, the crew cabin, and the nose cap. The data was examined by industry analysts in Houston, and fortunately no significant aberrations were detected. Everyone who was in any way involved heaved sighs of relief.

Atlantis docked at the International Space Station early on day three of the flight, as had been intended. Then, shortly after the hatches were opened, the six astronauts who had just arrived were warmly welcomed to the station itself. The crews had a brief reunion, and in no time were ready to begin the work that lay ahead. The first task was the transfer of some much-appreciated ancillary cargo to the station. That done, the all-important transfer of the huge P3/P4 truss began. Two astronauts on *Atlantis* used the Canadarm to lift the thing from the payload bay. Then Steve MacLean and Flight Engineer Jeff Williams, from the space station, used Canadarm2 to grapple the truss and position it for its ultimate destination as part of the station itself. In doing this, MacLean became the first Canadian to operate Canadarm2 in space. This procedure was truly a memorable accomplishment, and for that reason was mentioned with pride by several media outlets here on Earth.

The following day was what NASA called "Installation Day" at the International Space Station. Two American members of the *Atlantis* crew spent several hours spacewalking, and began the process of attaching the P3/P4 truss to an existing one truss P1. Large bolts held the two together, and again, MacLean and Williams used Canadarm2 to position the new truss so that the spacewalkers outside could make the connection. It was made, and the operation was done quickly and without incident. However, when a bolt and washer from another connection came loose, they floated away in space. Fortunately, NASA managers on Earth were certain that the missing bits of metal would pose no danger, and would simply burn up when they fell into the atmosphere of Earth.

Steve MacLean began his spacewalk at 4:05 a.m. CDT, on Wednesday, September 13, 2006. He and another walker, American Dan Burbank, spent just over seven hours outside, and despite setbacks and unexpected problems, finished the work they wanted to do. Much later, I asked MacLean what stepping into space meant to him.

"It is the most amazing thing," he explained. "The two dimensional window of the shuttle is nothing compared to the peripheral vision you have within your helmet outside. Your perceptions of speed, your perceptions of height and so on are far more accurate outside than they are inside. There is a world of difference between the two. But I won't say it's overwhelming, because if it had been overwhelming to me, I would have made a mistake in the first ten minutes. But I really focused so that I would not make a mistake at any time — but especially in those first few minutes. Being out there is an amazing."[14] As he was describing this most unique of experiences for me, I had the impression that this Canadian astronaut found it hard to put into words the true depth of meaning of what he had done. And no wonder. Very few human beings in the entire history of the world have been fortunate enough to have participated in such a thing.

That evening, after he and Burbank were safely inside again, NASA issued a status report describing the work of the two men. It summarized, in largely jargon-free terms, what had been done: "During the second spacewalk of the STS-115 mission," the statement began, "spacewalkers Dan Burbank and Steve MacLean devoted the day to the final tasks required for activation of the Solar Alpha Rotary Joint (SARJ). The SARJ is an automobile-sized joint that will allow the station's solar rays to turn and point toward the sun. Burbank and MacLean released locks that had held the joint secure during its launch to orbit onboard *Atlantis*. As they worked, the spacewalkers overcame several minor problems, including a malfunctioning helmet camera, a broken socket tool, and a stubborn bolt and a bolt that came loose from the mechanism designed to hold it captive. The stubborn bolt required the force of both spacewalkers to finally remove it."[15]

Here on Earth, media coverage was extensive and enthusiastic. "Canadian astronaut Steve MacLean and American Dan Burbank spent more than seven hours on the spacewalk, unpacking and testing a rotating arm for the power-generating solar array. The solar array will almost double the amount of electric power at the space station, bringing the National Aeronautics and Space Administration a step closer to finishing the half-built outpost by 2010. Reaching that goal is part of NASA's plan to retire the space shuttle fleet the same year and build a vehicle that can also reach the moon,"[16] was the way one newspaper described what the two spacewalkers had done.

Another focused on the visual experience of being outside:

> Orbiting 400 kilometres above the coastline of South America, with a glorious backdrop of cloud-speckled ocean scrolling slowly behind him, Steve MacLean hung upside down, clad in his white pressurized suit. For nearly seven hours yesterday morning, Mr. MacLean and Commander [Dan] Burbank, a U.S. Coast Guard officer, worked outside the International Space Station, readying a truss structure that carries giant solar panels. From the station's pressurized lab module, 30 metres away, the pair looked like small figures dangling on the sides of a towering building. There were also breathtaking sights of the two hanging next to each other while underneath their feet an ochre and green land mass slowly edged out of view into the faded blue of the sea.[17]

Sometime later, the same newspaper carried what amounted to a follow-up to the account:

> Mr. MacLean recalled the stunning sights he saw while orbiting above Earth at 25 times the speed of sound: the Great Lakes all visible in a single glimpse, the Maritime

provinces. He came out of the space station's airlock feet first, so his first sight when the hatch opened was a swath of blue and white ocean, scrolling 400 kilometres away, between the boots of his spacesuit. He spent the first five to seven minutes adjusting to his surroundings.[18]

The spacewalks by all four astronauts became notable segments in most television news broadcasts, with extraordinary snippets of them working on the truss installations, the scudding white clouds over the Earth as a backdrop. Talking heads on various programs described what was happening, generally making reference to the fact that this mission was the most important "since *Columbia.*" Those two words were enough for the television audience, who understood the meaning behind them and hoped and prayed that there would never be such an accident again.

A third, and final, spacewalk was conducted, and when it was successfully concluded, the work outside was done. All the installations had been made, and the giant solar arrays were spread, looking like "wings of gold," some said. Once the important tasks were finished, several media outlets did crew interviews, and there were the usual congratulatory calls from politicians. Then the last few supplies brought on the shuttle were transferred to the space station. Included in this drop-off were ninety pounds of oxygen. In turn, unneeded equipment and trash from the station was packaged and loaded for its return to Earth. In due course, crew goodbyes were said, the hatch between *Atlantis* and the station was closed, and the shuttle moved away from the largest man-made object on orbit. In all, there had been six days, two hours, and two minutes of joint operations between the two spaceships.

After *Atlantis* undocked, it was inspected carefully to insure there had been no damage to it from any source, from the time of blast-off to the moment it turned for home. Steve MacLean was instrumental in this examination, as were his colleagues, and Canadarm was used to perform the task. While no significant abnormality was noted on any part of the orbiter, something else was seen.

A chunk of unidentified debris was floating near the shuttle. The detection was made by shuttle managers in Houston, and when no one there could ascertain exactly what the thing was, the astronauts were alerted. At about the same time, Mission Control elected to extend the mission by at least one day — for two reasons: they wanted to do some further probing concerning the debris, and they noted that weather conditions in Florida would not be favourable for landing on the day planned. *Atlantis* had plenty of supplies on board for at least three or four extra days, so the continuation of the mission was not a worry. The debris was.

Back on Earth, reporters who covered the space programs picked up on the anomaly and wrote about it. One such account was rather alarmist: "NASA postponed the return of *Atlantis* for at least a day and examined the shuttle for damage that could prevent it from making the journey home after a mysterious object apparently fell off the ship," the story ran. "Space agency officials wanted extra time to establish whether the object was a vital piece of the shuttle — such as tiles that protect it from the blowtorch heat of re-entry — and whether it harmed the spacecraft when it fell away. Officials were not optimistic that they would be able to identify the object, since the possibilities were almost endless, ranging from harmless ice to crucial thermal protection tiles. But the leading candidate was a plastic space-filler placed between thermal tiles." The article went on to say that "the space agency did not rule out the possibility of a spacewalk to make repairs or, if the spacecraft is too damaged, sending *Atlantis*'s six crewmembers to take refuge in the international space station and await rescue by another shuttle." Then, almost in an afterthought, the piece mentioned that "the object floated near the shuttle in the same orbit for a while, slipping farther and farther away until it was just a dark speck in NASA video beamed down to earth."[19]

Ultimately, no missing parts were detected on the shuttle, but some damage from an unknown source was discovered later. Three weeks after *Atlantis* had safely landed, Steve MacLean took part in a news conference at the Canadian Space Agency. In an answer to a reporter's question, he mentioned that something, and no one knew exactly what, had caused a small puncture in a radiator on board. An examination of the orbiter had detected the problem, and it was felt that a micro-meteorite could have been the culprit. In any case, if a cooling tube on the spacecraft had been cut, Freon gas would have leaked out, and shuttle cooling could have been effected. "We were lucky," MacLean admitted, because the unknown object "hit a five inch gap in the cooling system where there are no cooling tubes."[20]

As the shuttle was heading home, no one was able to identify the debris, but neither it, nor the possible meteorite impact, made *Atlantis* unsafe. The vehicle was examined over and over again, and finally NASA declared that their engineers "found no issues of concern."[21] Mission Control gave the all-clear for the de-orbit burn; the spaceship endured the heat and flame of re-entry, and at 6:22:16 EDT on Thursday, September 21, the wheels of *Atlantis* stopped safely on Runway 33 at the Kennedy Space Center.

Four days later, the entire crew and their families, including Steve MacLean, his wife Nadine, and their teenagers Jean-Philippe, Catherine, and Michele, found themselves at the White House in Washington. President George W. Bush had called them there to say thanks.

Early in the morning on September 21, 2006, the space shuttle *Atlantis* touched down at Kennedy Space Center in Florida, after almost twelve days in space. Canadian astronaut Steve MacLean was a member of the crew. During the flight the spaceship travelled almost five million miles.

17

Build the Station, Build the Future

The first time I ever saw Dave Williams, he and the rest of a spaceship crew were walking out of the Operations and Checkout Building at the Kennedy Space Center, in Florida. They were clad in the familiar orange launch and entry spacesuits, and were smiling and waving at the assembled media. Two minutes later, they were gone. Almost as quickly, so were the reporters.

This "walkout" was part of a carefully choreographed series of steps that occur prior to each space shuttle launch. It is a chance for news people, and a few others, to catch a last glimpse of crew personnel as they board the Astrovan that takes them on the twenty-five minute drive to the launch pad being used that day. In Williams' case, the van headed for 39A, where the shuttle *Endeavour* waited. Once there, they would take the 195 foot elevator up to the White Room, next to the shuttle crew access hatch, and ultimately crawl into the orbiter and take their positions for blast-off. Gantry technicians would secure safety straps for each flier, all of whom were lying on their backs with their knees upward. The necessary communications and other hook-ups were made and checked, and, finally, the hatch was sealed as the countdown continued. Barring a scrub, no one would see the astronauts on Earth until their mission had ended.

On a personal note, I found the walk-out photo op somewhat disconcerting. It was like a last wave to those who were flying; a bittersweet goodbye until they returned. And, watching Dave Williams go, I could not help but wonder if I would ever see him again. I talked to reporters who were there, and some of them vividly recalled the walk-outs for both *Challenger* and *Columbia*. On those two terrible occasions, the goodbye waves were final ones, and neither would be forgotten.

Of course, I did see Dave Williams again. This time, it was in a small, tastefully neat office on the eighth floor of St. Joseph's Hospital in downtown Hamilton, Ontario. Now, instead of the flight suit, the flag patches, and the big boots, Doctor Williams wore freshly pressed slacks, a shirt and tie, and dress shoes. A firm handshake replaced the obligatory wave to the crowd on the day I had seen him last. Now, instead of an active astronaut about to fly, he was a professor of surgery in the Faculty of Health Sciences at McMaster University. It was good to get a chance to talk to the man. He is urbane, congenial, helpful, and approachable. He is also a raconteur with an unquenchable curiosity. Even though he would be loathe to admit it, he is undeniably one of Canada's modern heroes. I had come to talk about his career as an astronaut, in particular his second flight and the three spacewalks he did while on it.

Because I had been at Kennedy the day he left the Earth, and for a few days prior to the departure, I had at least a rudimentary understanding of the preparations, sequence of events, and countdown procedures that were involved. I had seen the orbiter as it stood waiting to go; I had gone to the pad the night before liftoff and admired the shuttle, gleaming and white in the powerful xenon lights, and I had heard, seen, and felt the majesty of the liftoff when the mission began. But this was the first time that I had met this man who was such an integral part of it all.

STS-118, as Doctor Williams' second and final flight was called, involved an important building project at the International Space Station. At the precise time that *Endeavour* and its crew of seven left Florida, the station with which it would dock a couple of days later was 212 miles

Courtesy John Melady

Canadian Doctor Dave Williams waves to reporters as he walks with Barbara Morgan to the Astrovan, which transports astronauts to the space shuttle they will fly in. Barbara Morgan was the backup to Christa McAuliffe, who lost her life when the *Challenger* exploded above the heads of thousands of horrified spectators at Cape Canaveral.

above the Atlantic, just southeast of Halifax. On board it were three men, one American and two Russians. When the *Endeavour* crew arrived, all ten individuals would work together to complete the tasks at hand. The spaceship en route was transporting three main items. In NASA parlance, these were a "4,000-pound Starboard 5 truss segment, a replacement for a Control Moment Gyroscope and a 6,900-pound External Stowage Platform, to stow replacement parts outside the station."[1]

The flight would be the twentieth for *Endeavour*, which first flew as STS-49 back in 1992. The orbiter was built as a replacement for *Challenger*, but for the four years prior to this Williams mission had been on the ground at Kennedy, being substantially upgraded. It was, and still is, NASA's newest shuttle, and the downtime was because of budget tightening and other factors. But "during its stay on earth, about 200 technicians performed routine maintenance work on the ship, outfitting it with thousands of new parts. *Endeavour* also received nearly 200 upgrades to its systems."[2] Included in the overall upgrade was the replacement of "2,045 of *Endeavour's* heat tiles,"[3] and hundreds of gap fillers between the tiles.

The launch of the orbiter was picture perfect, and its first hours on orbit were just as positive. A nagging worry cropped up soon though. A few bits and pieces of foam appeared to have come loose during lift-off, but likely late enough that it was felt they would not be a danger to the shuttle. So, with that reasoning, as the spaceship sped towards the space station, even NASA's all-inclusive moniker for the flight seemed guard-edly appropriate. It was dubbed: "Build the Station, Build the Future."

When he went into space, Dave Williams said goodbye to his wife of twenty-one years, Air Canada Pilot Cathy Fraser, and their children: Evan, twelve, and Olivia, nine. At the time, the family was living in a community not far from the Johnson Space Center, and while the children took an active interest in what their father was doing, it was not necessarily novel to them. That was because several parents of the children's schoolmates were astronauts, and many of them had done spacewalks as well.

"Around here, a spacewalk is pretty routine," Fraser told a reporter at the time, "because we are surrounded by people who have actually done that sort of thing." Nevertheless, the children were excited for their father. "They see other parents go off to space and come back two weeks later, and all the hoopla that goes with it,"[4] she added. But neither she nor the youngsters would be far from Williams' thoughts during the flight. He took personal mementoes of theirs with him, including a CD of his daughter playing "It's a Wonderful World" on the piano. He always loved the song, and Olivia's rendition of it added a delightful emotional depth for him.

There were five men and two women on the flight. As mentioned earlier, one of the latter was teacher–astronaut Barbara Morgan. She was the replacement for Christa McAuliffe, the original teacher-in-space, who perished on *Challenger*. The second woman on board was Tracy Caldwell, who holds a doctorate in chemistry and can converse in American Sign Language and Russian. She was the designated intravehicular officer on board, and as such, would play an integral role in the choreography for the spacewalks scheduled for the mission.

During their first full day in space, crewmembers used the Canadarm and boom to check out the spacecraft, to locate and assess any damage that falling foam, or even ice, on liftoff might have caused. The heat shield on the leading edges of the wings received particular attention. Technicians on the ground determined that "the first foam fragment came off at 24 seconds after liftoff and appeared to hit the tip of the body flap. The second was 58 seconds after liftoff, with a resulting spray or discoloration on the right wing. The third came almost three minutes after liftoff, too late to cause any damage to the wing. The most worrisome is one that appeared to hit the right wing."[5]

While these concerns were being studied at Mission Control, the spacesuits to be worn during the scheduled spacewalks were double checked by those who would wear them. The schedules for the walks were reviewed, as were the detailed objectives that would be carried out. Nothing, even the smallest detail, was left to chance.

The *Endeavour* had left the pad at 6:36 EDT on a Wednesday. By noon, two days later, it was preparing to dock at the space station. And, as has become the norm since *Columbia*, when the arriving shuttle came close to its destination, the orbiter was flown on a kind of back flip so that its underside could be photographed from the station. In this case, 296 digital photos of *Endeavour* were taken, all of which were immediately downlinked to Mission Control.

Unfortunately, when they were checked, an aberration was noted, and in short order it became the subject of several news accounts as the mission continued. Typical of these was a story that ran the day after the photos were taken. "A spray of ice during liftoff might have gouged the heat shield of shuttle *Endeavour*, which has carried Canadian astronaut Dave Williams to the International Space Station. NASA managers said yesterday that a white spot near one of the shuttle's landing gear traps is likely a seven-centimetre gash in a heat tile. A closer inspection will be conducted."[6]

Closer inspections were made, and for the next few days, technicians in Houston carried out tests on the materials involved, and tried to come up with the best, safest, and most practical solution to the problem tiles. As the examinations were being made on Earth, Dave Williams and his partner, Rick Mastracchio, were preparing to walk above the clouds. The two moved out of the station at 11:28 a.m. Central Time, on Saturday, August 11, and remained there for over six hours. I asked Dave Williams to tell me what doing a spacewalk meant to him.

"Well, to me, it is such an honour, and it's such a pleasure to do. On my first flight, STS-90, I was trained as a contingency spacewalker, which meant that I would have gone outside if there had been a problem. Fortunately, we did not have anything that went wrong, so I did not have to go. But you have mixed feelings about that," Williams added. "You hope to have the experience, but you don't want a problem. So, on my second mission, I was ready.

"To me, spacewalking is a true test of your ability to work as a member of a team, and I can't stress that enough. Rick and I did our work, but we relied heavily on Tracy Caldwell so that we could get things done. She was what is called the IV crew person, or the Intravehicular Officer, and she has to work seamlessly with the spacewalkers, doing the robotics to integrate everything. The IV person has to choreograph the whole thing, so that it happens in the most efficient manner possible. She did a great job of doing that, and Rick and I had to learn to work together, and work with Tracy to make this all happen.

"You know, when you are a spacewalker, you are kind of independent team player, which sounds like an oxymoron. But by that I mean that you have to be able to go forward on your own, troubleshoot problems, think about where you are on the timeline, make sure everything is moving forward, but at the same time, do that in conjunction with the team. So in some cases, you have to wait for the IV crew person to have you go forward to begin a task. Tracy, Rick and I really worked hard in training to achieve that, so by the time we were in space, we were able to go forward seamlessly to do what had to be done. It has been said that in spacewalking you should take your checklist and your procedures and throw them away as soon as you go out the hatch, because things change. But I will always think back to our working as a team, and how successful it was."[7]

And, as he mentioned, "things change," or at least can happen without warning when working in space, as they do here on Earth. An example of this occurred during Williams' third walk, when he was outside with astronaut Clay Anderson, one of the men who had been living on the space station prior to the arrival of *Endeavour*. I asked Dave Williams to explain what happened.

"Well, that day, Clay and I were outside, working in separate places, doing separate things. Suddenly, the fire alarm sounded! We are in the middle of an EVA and we get a fire alarm!

"Clay said something like, 'Oh, it's a fire alarm. It's a station fire alarm,' but he didn't react otherwise.

"Then Tracy said: 'Okay guys, just continue. I'll let you know what's happening.' So we didn't even pause. We just kept on working. A few minutes later, Tracy got back to us and told us it was a false alarm."

Another unexpected occurrence happened on an earlier spacewalk, and although Dave Williams was not involved, Rick Mastracchio was. That time, after Mastracchio had been outside for more than four hours, he noticed a tear in the thumb of his left glove. However, even though the rupture only penetrated two of the five layers of the glove material, the astronaut was ordered to go back inside immediately. Had the tear penetrated all of the layers the consequences would have been grave. This was because the entire suit that astronauts wear on EVAs is really a small spaceship, designed to protect its wearer from the vacuum of space. If the protective shell of the suit becomes punctured, its wearer would perish.

The first American woman to walk in space described the dangers of working in such a hostile environment. During her training for a spacewalk, Kathy Sullivan told of technicians who pumped all the air out of a vacuum chamber in order to demonstrate for her the contrast between the Earth's air pressure, and the lack it in space. When all the air was removed from the vacuum chamber, it did not look any different. However, a shallow pan of room temperature water had been left inside. As Sullivan later wrote: "As the air pressure in the chamber decreased, the water began to bubble and then boiled vigorously. Then, in one dramatic flash, the water making up the walls of the bubbles froze and dropped into the pan as a layer of slush." Her comment about what she had seen: "If a picture is worth a thousand words, this vivid demonstration of what would happen to all the fluids in your body if your suit depressurized in space was worth a million."[8] No wonder Rick Mastracchio was told to hustle back inside the space station when he noticed that his glove was torn. And, even though NASA informed the media about the incident at the time, they insisted that the astronaut "was in no danger and that the hole did not penetrate all the layers of his glove."[9]

Problems aside, the astronauts on *Endeavour* had come to the space station to continue the building of the edifice. The work was hard, tiring, and required every skill they possessed and practiced. They also prepared themselves mentally for whatever they might have to face, in an environment that is completely alien to human beings.

"When you go outside for the first time," Dave Williams explains, "you are alone in the universe. It is truly a unique feeling. The best way to compare the difference is to use the analogy of going down a highway in a car, versus driving down the same road on a motorcycle. On the space-walk, you are on the motorcycle, and you feel so much more in the open. You are more alone; you are in your own spacecraft, in your own little world. So, when you open the hatch for the first time, you ask yourself how you are going to react. You look down — or up — whichever way you want to think of it, but you are looking at the earth, over 440 kilometres away, recognizing that you are travelling ten kilometres every second, at Mach 25. It's an incredible sensation."[10]

To prepare for his exit from the spacecraft, Doctor Williams made use of a software package specifically designed to help with the spaceship egress. He would also sit down with his eyes closed, doing his best to imagine every step he would take once he was outside.

"When you leave the spacecraft," Doctor Williams continued, "there is a real risk that you will be overwhelmed by the view, by the magnitude of the openness. If that happens, there could be significant mission impact and everything you have to do would be delayed. So as a crew, we train to guard against this. That was why, before going out, I used mental imagery and virtual reality to prepare for what it was going to be like. Because we didn't know if it was going to be dark or light when we went out, Rick and I made sure that for the six and a half hours of our first spacewalk, we basically memorized the whole thing. We memorized where we would be putting our hands, from handrail to handrail.

"So the goal we had, working with Tracy, was to go into auto-pilot when we went out the hatch, so that we would seamlessly transition right into our trained mode. Anyway, when we went out the first time, it was night, so we moved into our trained mode right away. That was important, because the station is now so big; it was about 160 feet to our work site, so we had two 85-foot tethers with us. We had to put one 85-foot tether onto a handrail, and use the second, just to get to our work site.

"By the time we got that done, I was on the outer surface of the station, and looking down at the Earth just as the sun was coming up. I had a brief chance to enjoy an absolutely spectacular view of the world because I was in the foot restrain, far out on the starboard side. But, because we are so completely focused on our work during that first time out, there was only a brief time to take in the view. On the second and third spacewalks though, we took a bit more time to enjoy the experience, but even then, we made sure we did not get distracted.

"As a matter of fact, on the walk when we had the fire alarm, and about half an hour after it, Tracy said to us: 'Hey guys, look, we are flying

Courtesy NASA

This picture of the shuttle *Endeavour* docked at the International Space Station was taken by a crew member during the third spacewalk on STS-118. Both Canadarm and Canadarm2 can be seen in the photograph — with the Earth and space as backdrop.

over the Gulf of Mexico. Hurricane Dean is right below you. Have a look!' I looked down, straight down into the eye of the storm, and it was absolutely incredible! But unless Tracy had alerted us, we would have continued working, and would not even have noticed it. That is because spacewalking is not only a huge physical challenge; it is a big mental challenge as well. In order to remember the sequence, the choreography and so on, you can't afford to let your attention wander."[11]

The hurricane mentioned by Williams was an unexpected occurrence during this flight of *Endeavour*, and it impacted in the latter stages of the mission. However, in the first days in space, there were plenty of tasks at hand, and the spacewalkers had their work cut out for them. On their first spacewalk, Rick Mastracchio and Dave Williams installed a two ton, eleven-foot-long device called a spacer. It was attached to the station's truss, or central "backbone," and once in place meant that the truss itself was now 246 feet long. By way of comparison, the ice surface of a professional hockey rink is two hundred feet long. However, this great construction project is far above the clouds and spinning around the world faster than a bullet shot from a gun.

One of the main objectives of STS-118 was the replacement of one of the station's four big gyroscopes. This was necessary because it had failed the previous year, and as these things controlled the orientation of the station in space, all needed to function. During their second spacewalk Williams and Mastracchio installed the new gyroscope and stored the faulty one on the outside of the station. The failed machine would then be brought back to Earth on a later mission.

The installation proceeded without incident and the entire time outside amounted to six hours and twenty-eight minutes. As the two spacewalkers went about their tasks, there were the usual NASA photos taken of them doing so. One of the most dramatic of these shows Williams standing on a foot restraint, the 600 pound replacement gyroscope in his arms, with clouds over the Pacific Ocean visible far below, and the velvet blackness of space beyond [see cover photo]. Holding, moving, and certainly lifting such a weight would be virtually impossible on Earth. The illustration is certainly a vivid example of how things can be done in space, where gravity is neither a factor nor a constraint.

In addition to their work on the station and outside it, the crewmembers of *Endeavour* participated in media and other interviews from space. Sometimes these were with news organizations alone, sometimes they involved school children, and at other times politicians in astronauts' hometowns and elsewhere. Because she was the first teacher in space, Barbara Morgan received, and had to juggle, far more interview requests than she had time to handle, but Dave Williams not only helped her out, he took part in many as well. Some of these were for radio, while others involved televised downlinks. One of the latter, as mentioned earlier, was with the Discovery Center in Boise, Idaho, and had been chosen by NASA because Morgan had taught in that state. And even though Dave Williams was never a teacher there, he was asked to participate because of his easy broadcast manner, and because of his ability to connect with youngsters. There were questions about what it was like to be weightless, how the crew got fresh air on the shuttle, what planets and stars looked like when viewed from space, why people in space became taller, and more.

Another broadcast involved Canadian school children in the community of La Ronge, Saskatchewan. During that twenty minute session, Doctor Williams answered the questions put to him, and then talked to his audience about their role in the future, how they should learn to recognize the explorer within themselves, and how each might play an active role in whatever their generation did in space. Then, as one of the print outlets on Earth added: "Doctor Williams, who was born in Saskatoon, talked to the students about his Saskatchewan roots and reminded them that anything is possible if they work hard."[12] Hopefully many took his words to heart.

While the work at the space station progressed, and as supplies and equipment were off-loaded from the shuttle, much soul searching and testing was being done in Houston, with regard to analysis of the damage to *Endeavour* during launch. For hours, and ultimately days, technicians studied the problem, tried to replicate the conditions in laboratories, and then did detailed assessments of what they found. The photos of the underside of *Endeavour* were studied, enlarged, and studied again. Because each tile on the bottom of the ship was numbered, the exact location of the damaged area was ascertained. Then, to see how such tiles would react, individual ones on Earth were subjected to temperatures the spaceship would encounter on its fiery ride back through the atmosphere.

Initially, there were five areas under the ship that received attention, but soon four of them were found not to be of concern. However, the fifth, where a worrisome tile gash was located, became the main focus and engineers at Houston's arc jet facility began to do thermal analysis of tiles similar to the ones damaged. These tests included "subjecting intentionally damaged tiles to heat and pressure comparable to shuttle re-entry conditions. Depending on the results of the tests, spacewalkers could be called upon to repair the damage."[13] In this bulletin from Mission Control, the possibility of astronaut repair was first broached. Then, in no time, Dave Williams and Rick Mastracchio were being mentioned as the likely crew persons who would be called upon to do such a thing. They had been given the training to do so, and had the requisite skills.

By this time, news outlets that followed the progress of this space voyage began to pick up on the possible seriousness of the problem on the shuttle. "A close-up laser inspection by *Endeavour's* astronauts revealed that a nine-centimetre-long gouge penetrates all the way through thermal tiles on the shuttle's belly, and has NASA urgently calculating whether risky spacewalk repairs are needed,"[14] reported one organization. The story went on to mention that "the unevenly shaped gouge — which straddles two side-by-side tiles and a corner of a third — is nine centimetres long and five centimetres wide. Yesterday's inspection showed that the damage went all the way through the 2.5-centimetre-thick tiles, exposing the felt material sandwiched between the tiles and the shuttle's aluminum frame."[15]

Further coverage was equally alarmist: "NASA experts were evaluating whether astronauts should venture into the void of space to fix a deep gash on the shuttle *Endeavour's* belly before the crew returns home," was typical. The same piece continued: "NASA put together a team of engineers, astronauts and spacewalk gurus who would be able to suggest the best method to pick and perfect the best way to fix the gauge and avoid extensive post-flight repairs. The gouge is relatively small and the damage is benign enough for *Endeavour* to fly home safely. But part of it penetrates through the protective thermal tiles, leaving just a thin layer of coated felt over the space shuttle's aluminum frame to keep out the more than 2,000-degree heat of re-entry. To patch the gouge," the article concluded, "spacewalking astronauts would have to perch on the end of the shuttle's 30-metre robotic arm and extension boom, be maneuvered under the spacecraft, apply protective black paint and then squirt in a caulk-like goop."[16]

Fortunately, in the end, NASA managers elected to do nothing because their experts finally concluded that the damage was not severe enough to pose a threat on the return flight. With the decision, which came at the end of the day on Thursday, August 16, Dave Williams and Rick Mastracchio put aside their thoughts of having to do shuttle repairs. They had spent considerable time preparing for such an eventuality and felt they were ready to attempt the task if necessary. The final meeting on the matter in Houston lasted over five hours, and involved assessing a massive amount of technical information. Some of the data led to concerns that "the heat of re-entry could weaken the shuttle's aluminum frame," but mission managers felt that having astronauts do repairs would have added risk and "they did not want to attempt it unless absolutely necessary."[17]

With the question of the damaged tiles set aside, just one day later another matter would demand the close attention of the NASA decision makers. Hurricane Dean, which Dave Williams mentioned earlier, had formed in the Atlantic, and a possible track for it was across the Gulf of Mexico, with landfall along the Texas coast near Galveston, and in a direct line to the Johnson Space Center just southeast of Houston. If that were to occur, there were fears that Mission Control would have to shut down, and in so doing, effect the scheduled return of *Endeavour*. Indeed, if that came about a small emergency mission control team was prepared to leave for Florida to support the intended landing there.

The hurricane watch became an intense one, and all of the computer updates on its track were followed closely. The swirling winds of the storm were extremely powerful, enough to do significant damage both to shoreline communities and inland as well. The storm was said to be moving at seventeen miles an hour, so it took some time to determine where landfall might occur. In the end, it bypassed Houston, but because there were always fears that this might not happen, NASA decided to cut the STS-118 mission by one day. This was in order "to avoid any disruption to flight operations in case the formidable storm takes aim at Houston."[18] By bringing *Endeavour* home a day early, it would be back and on the ground before Mission Control had to close. "It would be irresponsible for us not to pay attention to this storm,"[19] LeRoy Cain, chairman of the mission management team said at the time.

While the decisions about their mission were being made in Houston, the *Endeavour* crew, in conjunction with the men on the space station, went about their work. Dave Williams did a third spacewalk with Clay Anderson. The two men spent five hours outside, two less than previously planned. This was because Mission Control needed the hatch to the station closed earlier than intended in order to allow for undocking and to facilitate the earlier return to Earth. This walk accomplished three main objectives: as summarized in the status report of the day, the astronauts "installed the External Wireless Instrumentation System antenna, attached a stand for the shuttle's robotic arm extension boom, and retrieved the two materials experiment containers to be brought home on the shuttle." Because of the abbreviated timeline, "cleaning up and securing debris shielding and moving a tool box to a more central location — were deferred to a future spacewalk."[20]

Endeavour undocked from the International Space Station at 6:56 on Sunday morning, August 19. Following another close inspection of the tile under the shuttle, the leading edges of the wings, and the nose cap, the spaceship headed for home. Commander Scott Kelly and the rest of his crew affirmed their faith in the decision made by NASA managers that the astronauts not do repairs. All felt secure in their belief that the flight back would be a safe one. And it was.

On the last day on orbit, Houston wakened the sleeping crewmembers with a musical rendition called "Flying," by the Canadian group the Long John Baldry Trio. The piece was dedicated specifically to Dave Williams, because on this flight, he surely had been flying. His three spacewalks set a Canadian record, a total of seventeen hours and forty-seven minutes outside the space station. During the mission, on day five, he had also found time to participate in scientific experiments with another crew member. In sum, his contribution to the success of STS-118 was considerable.

As one newspaper reported, at 11:52 in the morning on August 21, 2007, as it was flying over the Indian Ocean,

> *Endeavour* fired its maneuvering thrusters for 3 minutes and 35 seconds. For the next hour, it dropped out of orbit towards Earth's atmosphere. The craft cruised over Costa Rica and Cuba, then over southern Florida, where the skies were blue with broken clouds and steady breeze.
>
> Two quick, loud sonic booms thundered across the space center as the shuttle passed to the east. It made a U turn for its final approach from the north. In a picture-perfect landing, its rear tires touched the runway at 12:32 p.m., and the nose gently swiveled to the ground 13 seconds later. Slowed by a drag chute, it rolled to a stop a minute later.[21]

After it was all over, I sat down to talk with Dave Williams about the mission, and about his illustrious career as an astronaut. I asked him if being in space changed him.

"Yes, I think so," he told me, "being in space helped me with a couple of things. You know, I've been a neuroscientist. I've been a medical doctor, an emergency trauma physician, an astronaut and an underwater explorer — an aquanaut. Now, when I look at that list, I finally understand a common theme: I'm a scientific explorer. For me, exploration is the quest for knowledge. It's truly that simple. I'm just one of those people who are truly curious about a lot of things. I like to understand physical phenomena, whether it's neuroscience, or taking apart the lawn mover engine or flying in space or being under water. This mission brought all these things together.

"Being in space was very humbling. I remember during the second spacewalk, riding on the end of the Canadarm, and feeling incredibly proud as a Canadian, and proud of the amazing team of individuals from Canada who were responsible for the Canadarm technology, and the achievements of the Canadian Space Agency. Then I would look down at

Canadian Mission Specialist, and veteran of three spacewalks, Doctor Dave Williams is now professor of surgery at McMaster University in Hamilton, Ontario.

Courtesy John Melady

the Earth, so far away from me, and think that that is the planet on which the entire history of the human species has taken place.

"But then, I would look out into space, into the infinite black void of space, this black like nothing you have ever seen in your life, and then, back at the earth, and realize how amazing it all was. Out there is so vast, so infinite, and apparently so hostile, while beneath me is the planet where everything that has happened to me; everything that has happened to the human species has taken place. All my loved ones are there. It was so remarkable, and so humbling."[22]

And with those few words, Dave Williams — explorer — went back to his first love: the healing of human beings.

18

The Forever Frontier

During a press conference held at the Canadian Space Agency in Montreal on August 18, 2007, Steve MacLean made several pertinent remarks, but two stood out. At the time, Dave Williams was part way through his record setting third spacewalk and reporters were asking for MacLean's reactions to the feat. He did not disappoint. He mentioned the physical requirements for spacewalking, but stressed that the mental ones were even more taxing. He said he was somewhat envious of Williams and what he was doing, but he also reminded his audience of the dangers that were involved. "If you make a mistake up there," MacLean stressed, "it will be your last mistake."[1] Then he predicted that in time, and despite the dangers involved, Canadians would set new records, and would accomplish even more in this exciting and largely unknown environment in which they worked.

This is already coming about. As this is being written, Drew Feustel, a personable young man with dual American-Canadian citizenship and a Ph.D. from Queen's University in Kingston, has just completed a shuttle mission on *Atlantis* that linked up with, and did upgrades to, the Hubble Telescope, that marvelous piece of machinery that has be in orbit around the Earth since 1990. Next, Julie Payette will be returning to space, this time on STS-127, scheduled for launch in early summer of 2009. On Wednesday, May 29, 2009, Bob Thirsk blasted off the steppes of Kazakhstan in a Soyuz rocket, to become the first person from Canada to live on the International Space Station. He will be there for six months. All three of these astronauts will play important roles in the human understanding and exploration of space. All of their missions are significant ones, and as Steve MacLean has stressed, all embody certain risks that will never be completely eliminated.

Feustel took part in something that, not so long ago, was regarded as too dangerous. Yet his was part of the fifth mission to Hubble, even though, at one point, upgrading the telescope was out of the question. In fact, "on January 16, 2004, NASA Administrator Sean O'Keefe announced his decision to cancel all further servicing missions to Hubble. Safety guidelines following the *Columbia* tragedy were cited as the primary basis for

Drew Feustel has Canadian/American citizenship, and a doctorate in geological sciences specializing in seismology, from Queen's University in Kingston, Ontario. His first mission in space involved a dangerous mission to service the Hubble Telescope.

Courtesy NASA

this decision."[2] However, NASA officials were finally convinced that one last mission should be attempted, despite the risks involved.

On Feustel's flight, if there had been problems *Atlantis* would not have been able to reach the International Space Station and temporary safety, so as a contingency plan if there was a crisis, "another shuttle would stand by ready on Kennedy Space Center's Pad 39-B. If it had been needed, *Endeavour* would have flown to Hubble and relieved *Atlantis'* crew within days."[3] While all this was inherently dangerous, that is one factor the astronauts played down.

"I really have no reservations about the plan," Drew Feustel told me when he was training for the mission. "I think it is as safe as they can make it. We understand the risks and we accept them. If you don't accept them; you don't fly. It's as simple as that."

In fact, some time ago, Drew Feustel talked about the Hubble and explained his position to his two young Canadian-born sons. "I told them what I was doing, and that there was risk involved. I also told them I could go up in the shuttle and not come back, but that I was going because I believed strongly in the importance of this flight — and in what I was doing — and that they needed to know that." Then this devoted father added: "I also told them that I hoped that in their lives they would find something just as important that they were committed to, and that would mean as much to them."[4]

Undoubtedly, the servicing mission was justified because Hubble has made unprecedented contributions to our understanding of the universe. Though a lengthy explanation of its importance is beyond the scope of this book, suffice to say that since the bus-size telescope was placed in orbit 304 miles above the Earth, it has circled the globe over one hundred thousand times, and has provided "more than 4,000 astronomers access to the stars not possible from inside Earth's atmosphere. Hubble has helped answer some of science's key questions and has provided images that have awed and inspired the world."[5] For those reasons, and so many more, keeping it functioning is fundamental to the expansion of knowledge of what is beyond our own planet and its closest neighbours. The servicing of the telescope will add at least five years to its operational time, likely much more than that: the Canadarm was used to facilitate the

Doctor Drew Feustel was a Mission Specialist for the dangerous mission to refurbish the Hubble Telescope.

Courtesy John Melady

work, Drew Feustel did three spacewalks on the mission, and Canadian astronomers will be among those who benefit.

The second Canadian flying, Julie Payette, will be exposed to all the usual dangers involved with space exploration. Despite having flown before, and having accepted the risks involved, neither she nor her fellow crewmembers will really relax until they are back on Earth. On STS-127 they will, in the words of the descriptor provided by the Canadian Space Agency, "install the Japanese Experiment Module Exposed Facility and Experiment Logistics Module Exposed Section."[6] Once complete, "the facility will provide a platform for experiments exposed to the space environment."[7] The Mission will last sixteen days, and involve five spacewalks. Payette's experience and expertise will greatly add to the potential of the mission. And if all goes well, she will join Bob Thirsk at the

International Space Station. The first time two Canadians will be in space at the same time.

Bob Thirsk flew to the International Space Station on board a Soyuz capsule. The spacecraft was launched by the Russian Space Agency, from their long-established facility at Baikonur, Kazakhstan. Doctor Thirsk's learning regime for the mission included studying the Russian system, learning Russian, and long training sessions on the Soyuz itself, with cosmonaut spacesuits and equipment. As anyone who has seen the Soyuz up close could attest, familiarizing himself with it took time. In fact, while I was researching this book in Houston, I had the chance to examine the Soyuz firsthand, and I marvelled at how anyone could fly in the thing at all. Its interior is unbelievably cramped, utilitarian, and devoid of almost every creature comfort. Yet, Bob Thirsk is excited about the mission, and about its potential as a tool for the advancement of knowledge.

Courtesy John Melady

Astronaut Bob Thirsk as he is today. In 2009 he became the first Canadian in history to live and work on the International Space Station.

That advancement of knowledge, and the desire to learn more about our world, is what has driven not only Canada's space program, but similar efforts in several countries. International cooperation in space has been achieved in ways that could not have been envisioned in decades past. The most obvious example is the amicable working relationship between Russia and the United States. Their programs, while not integrated completely, are much intertwined and are, in many ways, mutually dependent. Russians train and fly in the United States, and Americans do likewise in Russia, working shoulder to shoulder with their former Cold War antagonists. As well, both spend months and months together on the Space Station, "this great international, technological, and political achievement."[8] In fact, the station itself could be described as a kind of ultimate model of co-operation and accomplishment. Currently, sixteen nations, Canada among them, have had significant roles in the building and functioning of the station, and have invested in its potential and its future. For that reason, the results of work performed on the ISS "may be applied in various areas of science, enable us to improve life on this planet, and give us the increased understanding that can eventually equip us to journey to other worlds."[9]

Just as the Canadarm was our ticket to fly on the shuttle, back when Marc Garneau first flew twenty-five years ago, now our contributions, monetarily and otherwise, have led to Bob Thirsk's invitation to be one of the long-term space station personnel. In fact, as one journalist reported as early as 1987, "earning a place for Canadians aboard the station" has been an objective "since the inception of the Canadian astronaut program."[10]

Canada's role has been significant. Canadarm2, the space crane, and its work platform, the Mobile Base System, both of which are permanently attached to the station, perform operations that are critical to the functioning of this great, man-made satellite that is as unique as any structure ever built. Subsequent to the launch of Canadarm2 and its base system, another Canadian product, called Dextre, was added in 2008. It is the third and final component of the Canadian Mobile Servicing System, or MSS, and is a necessary tool for the external maintenance at the station. "Dextre," named by a contest for Canadian school children, is short for "dexterous," which is indicative of the skills of the two-armed, twelve foot robot. In fact, its "advanced stabilization and handling capabilities enable it to perform delicate human-scale tasks like removing and replacing small exterior components of the Station. Operated by crew inside the Space Station, or by flight controllers on the ground, it is also equipped with lights, video equipment, a stowage platform, and three different robotic tools."[11] Sometimes called the space station's robot, the machine cost over $200 million, and "is designed to assist spacewalking astronauts and, eventually take over some of their chores. Its designers envision the robot one day replacing batteries and other space station parts — it can lift up to 1,300 pounds — and also performing some fine precision tasks like handling bolts."[12] When I first saw Dextre, spread out on the floor of a building at the Kennedy Space Center, I found it almost impossible to envision the machine's ultimate function. So did many of the reporters who were there to hear scientific experts describe their creation and extol its capabilities.

But these additions to the space station are but a fraction of the Canadian involvement in space in recent years. They are detailed here because

Courtesy John Melady

The Special Purpose Manipulator, called Dextre, shortly before it was launched into space. Dextre was made in Canada, and is essential for maintaining and servicing the International Space Station.

they fall within the intent and parameters of this book. Other things Canadian, such as the weather station on the Phoenix Mars Lander and the Radarsat-2 satellite, which is monitoring the Northwest Passage, are equally dramatic, but beyond the scope of this narrative. Mention should be made, however, of the proposed sale of the manufacturing company, MacDonald Dettwiler and Associates (MDA), to the United States in the spring of 2008. Because Radarsat-2 was "developed in large part with a guarantee of $445 million federal contract for satellite data," and "Canada's only eye-in-the-sky capable of spotting a U.S. sub, for example, sailing through the Northwest Passage,"[13] Canadian reaction to the possible sale was generally negative. The fact that the United States does not recognize Canada's Arctic claims made the proposed "sell out" to the Americans quite controversial. In the end, after much public wrangling, the proposed deal was rejected by the Federal government, on the grounds that there was no "net benefit" to Canada if the sale went through. The decision received widespread support.

Inherent in the public controversy over the possible MDA sale was the ever-present financial concern over spending for projects in space. The overriding factor is always the expense, in spite of the vast array of benefits that have come out of space programs and their ancillary needs, including many widely used products and services, scientific, medical, communications, and other advances.

Indeed, there has been so much progress in human knowledge, skills, and accomplishments resulting from space-related work that it is almost impossible to enumerate or even comprehend it all. Perhaps the most obvious is in communications. Television, radio, cell phones, and the internet all function as they do because we have satellites that circle our globe, and human beings with the knowledge, skills, and daring to position these devices. We could scarcely function in the twenty-first century without the electronic links that encompass our world. Because of the multi-purpose and multi-faceted "eyes in the skies," we have military-related capabilities today that no previous generations ever knew. Global and regional mapping for a vast array of purposes is a fact of our existence — so much so that our everyday uses were not even dreams in ages past. Because of highly advanced inter-communications and robotic capabilities, we have made it possible for life sustaining surgeries to be performed, even when hundreds of miles exist between the physician and the scalpel being used.

Then there are the almost unlimited number of inventions and refinements of inventions that have come about because human beings had specific needs and expectations as they left the good Earth and journeyed beyond it. We have clothing made of new materials, tools that are job specific, life saving machines, improved sports equipment, better fire retardants, agricultural field mapping capabilities, mining inventions, search and rescue homing devices, and manufacturing refinements. All of them are taken for granted today, but they were invented because they were necessary for space travel.

A glimpse of capabilities in a variety of other areas: for solar energy, a new paper-thin plastic film absorbs sunlight and produces usable energy; ocean and forest imaging can provide data on everything from insect infestation to global warming; water purified through advanced filtration and treatment techniques can now be made available anywhere; oil spills can be better contained because of NASA technology using tiny balls of beeswax with hollow centers that absorb oil; hurricane and typhoon threats can be prepared for because of state of the art weather forecasting; better wind tunnel technology led to better aircraft design; new and better cameras exist because astronauts demanded them; the more mundane — better sunglasses are being sold today because their prototypes were needed in space; and a highly specific use for a related kind of lens, one that "provides dentists protection against the ultraviolet light they are exposed to when operating curing systems for dental materials."[14] In essence, there are few areas of human endeavour, great or small, that have not been touched in some way by space-related inventions and advances.

So where do we go from here?

Government assistance for, and treatment of, the Canadian Space Agency and its initiatives has been lacking for some time. Fortunately, that may be about to change. A third astronaut program was launched in the spring of 2008, and two new fliers were named out of 5,351 hopefuls who applied. They are Jeremy Hansen, a CF-18 fighter pilot who was born in London, Ontario, and David Saint-Jacques, a medical doctor who was born in Quebec City. Also in 2008, Steve MacLean was appointed head of the Canadian Space Agency. Undoubtedly these are positive moves.

Hopefully, by building on our past successes, and with the new personnel and new leadership, Canada will achieve much in space in this century; exploring the forever frontier, the limitless universe that we neither know nor understand. Because humans have already been to the moon, perhaps there is a child in this country who will one day leave the Earth behind and become the first Canadian to walk on the dusty surface of Mars.

NOTES

Chapter 1: A Pillar of Fire, Blinding and White

1. Jay Barbree, Conversation with the author, August 6, 2007.
2. Jay Barbree, *Live from Cape Canaveral* (New York: HarperCollins, 2007), 232.
3. James R. Hansen, *First Man* (New York: Simon & Schuster, 2005), 421.
4. Knowlton Nash, *History on the Run* (Toronto: McClelland & Stewart, 1984), 128.
5. Michael Griffin, "Bittersweet Triumph for 1st Teacher in Space," *USA Today*, August 9, 2007.
6. Lisa Stark, Conversation with the author, August 8, 2007.

Chapter 2: Crumpled Metal, Wires, and Waste

1. *The Canadian Oxford English Dictionary* (Toronto: Oxford University Press, 2004).
2. Chris Gainor, *Canada in Space* (Edmonton: Folklore, 2006), 17.
3. Philip de Ste. Croix, Ed., *Space Technology* (New York: Salamander Books, 1989), 253.
4. *Ibid.*, 254.
5. Chris Gainor, *Canada in Space* (Edmonton: Folklore, 2006), 25–26.
6. Lynda Hurst, "50 Years Ago, Tiny Sputnik Launched the Space Age," *Toronto Star*, October 4, 2007, 3.
7. Editorial, "The Legacy of Sputnik," *New York Times*, October 4, 2007.
8. William E. Burrows, *This New Ocean* (New York: Modern Library Paperback Edition, 1999), 188.
9. *Ibid.*, 181.
10. *Ibid.*, 29.
11. Andrew Smith, *Moondust* (New York: HarperCollins, 2005), 168.
12. Piers Bizony, *Space 50* (New York: Smithsonian Books, 2006), 40.
13. Tom Wolfe, *The Right Stuff* (New York: Bantam, 1979), 56.

Chapter 3: Trajectory of Achievement

1. Chris Kraft, *Flight* (New York: Penguin, 2001), 1.
2. William E. Burrows, *This New Ocean* (New York: Modern Library Paperback Edition, 1999), 212.
3. Philip de Ste. Croix, Ed., *Space Technology* (New York: Salamander Books, 1989), 261.
4. Tom Wolfe, *The Right Stuff* (New York: Bantam, 1979), 175.
5. Liberty Bell 7 remained on the ocean floor for thirty-eight years. It was successfully recovered in July 1999, and ultimately put on public display for the first time in 2006.
6. The *Randolph* (CV-15) was a ship with a storied past. Commissioned in October, 1944, she saw action in the Pacific during the Second

World War, and on March 11, 1945, lost twenty-five men as the result of a kamikaze attack on her deck. Recommissioned in 1953, the 888-foot-long ship served in the Atlantic and Mediterranean until 1969. Following his historic three-orbit flight in February 1962, astronaut John Glenn was, like Grissom, flown to the *Randolph*. The author toured the carrier during its time in the Atlantic.

7. Chris Gainor, *Canada in Space* (Edmonton: Folklore, 2006), 28.

Chapter 4: Canada's Membership Ticket

1. Pierre Salinger, *John F. Kennedy: Commander in Chief* (New York: Penguin, 1997), 98.
2. John Melady, *Pilots* (Toronto: McClelland & Stewart, 1989), 234.
3. Chris Kraft, *Flight* (New York: Penguin, 2001), 87.
4. Peter Margolin, "The Making of Lunar Explorers," *Geotimes*, August 2000.
5. "Canadian Space Milestones," Canadian Space Agency, 1973.
6. When I was being shown the Canadarm at the training facility at the Johnson Space Center, a NASA official mentioned the name variation, but because I am Canadian, was careful to use "Canadarm" in my presence.
7. Chris Hadfield, Conversation with the author, August 26, 2006.
8. Lydia Dotto, *Canada in Space* (Toronto: Irwin, 1987), 43.

Chapter 5: A Flash, a Roar, and a First

1. Lydia Dotto, *Canada in Space* (Toronto: Irwin, 1987), 36.
2. *Ibid.*, 97.
3. "PM May Talk to Canada's First Astronaut During Flight," Canadian Press, October 3, 1984.
4. Andrew Chaikin, *A Man on the Moon* (New York: Penguin, 1998), 288.
5. Lydia Dotto, *Canada in Space* (Toronto: Irwin, 1987), 17.

6. "Space 'Voyage of a Lifetime' Ends for Canada's Garneau," *Toronto Star*, October 14, 1984.
7. *Ibid.*
8. "Scared and Excited Astronaut Garneau 'Is Proud to Be Canadian'," *Toronto Star*, October 10, 1984.
9. Editorial, "Flying High for Canada," *Toronto Star*, October 4, 1984.
10. Over the years, shuttles have landed at three places: Edwards Air Force Base in California; the Northrup Strip in White Sands, New Mexico; and Cape Canaveral, Florida.

Chapter 6: A Moment of Hope and Pride

1. Michael Gorn, *NASA: The Complete Illustrated History* (New York: Merrell, 2005), 222.
2. William E. Burrows, *This New Ocean* (New York: Modern Library Paperback Edition, 1999), 556.
3. Michael Cabbage and William Harwood, *Comm Check* (New York: Simon & Schuster, 2004), 262.
4. Andrew Chaikin, *A Man on the Moon* (New York: Penguin, 1998), 573–74.
5. Andrew Smith, *Moondust* (New York: HarperCollins, 2005), 294.
6. Robert Freitag, "When the Space Shuttle Finally Flies," *National Geographic*, March 1981, 317.
7. Barbara Bondar and Dr. Roberta Bondar, *On the Shuttle* (Toronto: Greey de Pencier, 1993), 32.
8. Roberta Bondar, Speaking at The Green Living Show, Toronto, April 29, 2007.
9. Lydia Dotto, *Canada in Space* (Toronto: Irwin, 1987), 22.
10. Joan Dixon, *Roberta Bondar* (Canmore: Altitude, 2004), 86.
11. Editorial, "Dr. Roberta Bondar," *Toronto Star*, January 23, 1992.
12. Roberta Bondar, *Touching the Earth* (Toronto: Key Porter, 1994), 24.
13. *Ibid.*, 48.
14. Alan Mortimer, "Bondar Spends Hour Tidying Up Shuttle," *Toronto Star*, January 23, 1992.

15. *Ibid.*

16. Rod Goodman, "Astronaut Was Seen as Maid," *Toronto Star*, January 25, 1992.

17. *Ibid.*

18. Roberta Bondar, *Touching the Earth* (Toronto: Key Porter, 1994), 76–77.

Chapter 7: No Longer a Trainee

1. Tom Wolfe, *The Right Stuff* (New York: Bantam, 1979), 312.

2. NASA Facts, "Space Shuttle Transoceanic Abort Landing (TAL) Sites," 2006, 2.

3. Lydia Dotto, *The Astronauts: Canada's Voyageurs in Space* (Toronto: Stoddart, 1993), 124.

4. Chris Gainor, *Canada in Space* (Edmonton: Folklore, 2006), 178.

5. Rob Robichaud, "Blast-Off Blast," *Ottawa Citizen*, October 22, 1992.

6. Nadine Wielgopolski, "Hold Your Breath, It's Blast-Off," *Ottawa Citizen*, October 23, 1992.

7. Paul McLean (*sic*), "Canadian Marvels at 1st Taste of Space," *Toronto Star*, October 23, 1992.

8. Michael Smith, "Canadian Astronaut Set for Busy 10 Days in Space," *Toronto Star*, October 22, 1992.

9. Steve MacLean, "Spaceman Steve," *Ottawa Citizen*, October 27, 1992.

10. Lydia Dotto, *Canada in Space* (Toronto: Irwin, 1987), 137.

11. Nadine Wielgopolski, "Going in Circles," *Ottawa Citizen*, November 1, 1992.

Chapter 8: Our Top Gun in Space

1. Chris Hadfield, Conversation with the author, August 6, 2007.

2. Chris Hadfield, "Touching the Future," *Airforce*, April 1993, 25.

3.   Joe Vick, Conversation with the author, March 5, 2008.

4.   Mensa website.

5.   Carl Mollins, "Chris Hadfield," *Maclean's*, December 26, 1994, 65.

6.   Tom Wolfe, *The Right Stuff* (New York: Bantam, 1979), 17.

7.   Chris Hadfield, Conversation with the author, August 6, 2007.

8.   Linda K. Glover et al., *National Geographic Encyclopedia of Space* (Washington: National Geographic, 2005).

9.   *Ibid.*

10. Helene Hadfield, "Canadian Blasts Off into Historic Shuttle Role," *Toronto Star*, November 13, 1995.

11. Robert Godwin, *Space Shuttle Fact Archive* (Burlington: Apogee, 2007), 18–19.

12. Chris Hadfield, "Shuttle Set for Assembly Job," *Toronto Star*, November 14, 1995.

13. Chris Hadfield, "They're Sky-High with Success," *Toronto Star*, November 16, 1995.

Chapter 9: Fear Comes in the Months Before

1.   Bob Thirsk, Conversation with the author, September 28, 2007.

2.   Lydia Dotto, *Canada in Space* (Toronto: Irwin, 1987), 66.

3.   Conversation with the author, September 28, 2007.

4.   Andrew Smith, *Moondust* (New York: HarperCollins, 2005), 1.

5.   *Ibid.*, 3.

6.   Bob Thirsk, Conversation with the author, September 28, 2007.

7.   Stan Fischler, Dan Diamond, ed., *Total Hockey* (Kansas City: Andrews McMeel, 1998), 164.

8.   STS-78, "Day 12 Highlights," NASA.

9.   Lori Thorlakson, *Vancouver Sun*, June 21, 1996.

10. *Vancouver Sun*, June 25, 1996.

Chapter 10: You're Going into Space, or You're Going to Blow Up

1. Bjarni Tryggvason, Conversation with the author, April 9, 2008.
2. *Ibid.*
3. *Ibid.*
4. Tom Spears, "Blast-Off! *Discovery* Hurtles Sixth Canadian into Space," *Toronto Star*, August 8, 1997.
5. "Shuttle Finishes 'Perfect' Mission," Associated Press, August 20, 1997.
6. "Canadian Astronaut Is Western's Engineer-in-Residence," *Western Alumni Gazette*, Spring 2006.

Chapter 11: Under the Ocean and over the Clouds

1. Andrew Phillips, "Ready for Blast-Off," *Maclean's*, April 13, 1998.
2. Lydia Dotto, *The Astronauts: Canada's Voyageurs in Space* (Toronto: Stoddard, 1993), 4.
3. *Ibid.*, 4.
4. Amiko Nevills, "Preflight Interview: Dave Williams," NASA, July 7, 2007.
5. Canadian Press, "Canadian Doctor Blasts into Space on Shuttle," The *Leader-Post*, April 18, 1998.
6. Amiko Nevills, "Preflight Interview: Dave Williams," NASA, July 7, 2007.
7. Joseph Hall, "Canadian Lifts Off on *Columbia*," *Toronto Star*, April 18, 1998.
8. "Rats Go Under the Knife — in Space," *Toronto Star*, April 19, 1998.
9. *Ibid.*
10. Joseph Hall, "Ground Control to Dr. Dave: Students Talk to Space Shuttle," *Toronto Star*, April 26, 1998.
11. "Canadian Astronaut Ready," The *Leader-Post*, April 16, 1998.
12. Shawn McCarthy, "Scuba Diving Doctor Next Canadian in Space," *Toronto Star*, August 13, 1996.

13. "Shuttle Returns, Brain Research Continues," The *Leader-Post*, May 4, 1998.

14. Amiko Nevills, "Preflight Interview: Dave Williams," NASA, July 7, 2007.

15. *Ibid.*

Chapter 12: Building a Cruise Ship in the Middle of the Ocean in a Storm

1. Julie Payette, *Julie Payette, Canadian Astronaut* (Don Mills: Pearson Education Canada, 2005), 7.

2. Julie Payette, Conversation with the author, August 26, 2006.

3. Julie Payette, Conversation with the author, August 8, 2007.

4. Lydia Dotto, *The Astronauts: Canada's Voyageurs in Space* (Toronto: Stoddart, 1993), 16.

5. *Ibid.*, 7.

6. Julie Payette, "Ambition Accomplished," *Maclean's*, May 17, 1999.

7. Andrew Phillips, "Out of This World," *Maclean's*, May 17, 1999.

8. Julie Payette, "Ambition Accomplished," *Maclean's*, May 17, 1999.

9. Chris Wattie, "Happy Family Sees Payette Off into Space," *Montreal Gazette*, May 28, 1999.

10. Mark Abley, "College Reveres Alumna Payette," *Montreal Gazette*, May 28, 1999.

11. "Payette Calls Shots on Stroll in Space," *Montreal Gazette*, May 30, 1999.

12. Edison Stewart, "Children, PM Hook Up with Astronaut," *Toronto Star*, June 2, 1999.

13. Jack Branswell, "I'm Ready to Go Back: Payette," *Montreal Gazette*, June 8, 1999.

Chapter 13: Like Christmas Eve When You Are Seven

1.   Lydia Dotto, *The Astronauts: Canada's Voyageurs in Space* (Toronto: Stoddart, 1993), 55.
2.   Robert Godwin, *Space Shuttle Fact Archive* (Burlington: Apogee, 2007), 30.
3.   *Orlando Sentinel*, August 8, 2007.
4.   Lydia Dotto, *The Astronauts: Canada's Voyageurs in Space* (Toronto: Stoddart, 1993), 39.
5.   *Ibid.*
6.   Mark Nichols, "The Selling of Space," *Maclean's*, May 27, 1996, 46.
7.   *Ibid.*
8.   Stephen Strauss, "Garneau's Second Shuttle Trip a Down-to-Earth Affair," *Globe and Mail*, May 18, 1996.
9.   Joseph Hall, "Our Man in Space … Again," *Toronto Star*, May 19, 1996.
10. *Ibid.*
11. Nicolaas Van Rijn, "Metro Students Visit Blastoff," *Toronto Star*, May 20, 1996.
12. *Ibid.*
13. STS-77 Mission Control Center, "Status Report 1," May 19, 1996.
14. Marc Garneau, "Garneau's Second Flight into History," *Maclean's*, June 3, 1996.
15. STS-77 Mission Control Center, "Status Report 10," May 23, 1996.
16. STS-77 Mission Control Center, "Status Report 21", May 29, 1996.

Chapter 14: Waving the Canadian Flag at the World

1.   "Space Shuttle Liftoff Launches 'Serene' Canadian Astronaut," *Toronto Star*, December 1, 2000.
2.   "Garneau Wings Way into Space," *Toronto Star*, December 2, 2000.
3.   STS-97 Mission Control Center, "Status Report 5," December 2, 2000.
4.   "Solar Wings Unfurl in Space," *Toronto Star*, December 4, 2000.

5. Colin Nickerson, "From Space Interloper to 'Part of the Program'," *Toronto Star*, December 11, 2000.

6. *Ibid.*

7. "Garneau Links Up with Youngsters," *Toronto Star*, December 10, 2000.

8. Conway Daly, "Garneau's Final Frontier," *Toronto Star*, December 9, 2000.

Chapter 15: The Most Memorable Thing an Astronaut Can Do

1. Johnson Space Center News Release, May 15, 1997.

2. Chris Gainor, *Canada in Space* (Edmonton: Folklore, 2006), 224.

3. Chris Hadfield, Conversation with the author, August 6, 2007.

4. Andrew Chung, "Canada's 'Down to Earth' Astronaut," *Toronto Star*, April 19, 2001.

5. Chris Hadfield, Conversation with the author, August 6, 2007.

6. *Ibid.*

7. Michelle MacAfee, "Hadfield Spacewalk Giant Step for Canada," *Toronto Star*, April 23, 2001.

8. Andrew Chung, "They're Floating on Air," *Toronto Star*, April 23, 2001.

9. STS-100 Mission Control Center, "Status Report 11," April 24, 2001.

10. STS-100 Mission Control Center, "Status Report 20," April 28, 2001.

11. "High Drama for Endeavour," *Toronto Star*, April 21, 2001.

Chapter 16: Storms, Malfunctions, and Risk

1. Michael Cabbage and William Harwood, *Comm Check...: The Final Flight of Shuttle Columbia* (New York: Simon & Schuster, 2004), 2.

2. Michael Gorn, *NASA: The Complete Illustrated History* (New York: Merrell, 2005), 274.

3. Michael Cabbage and William Harwood, *Comm Check...: The*

Final Flight of Shuttle Columbia (New York: Simon & Schuster, 2004), 248.

4. Michael Gorn, *NASA: The Complete Illustrated History* (New York: Merrell, 2005), 276.

5. *Columbia* Accident Investigation Board, Volume 1, 49.

6. Dennis R. Jenkins and Jorge R. Frank, *Return to Flight* (North Branch, MN: Specialty Press, 2006), 110.

7. Todd Halvorson, "Monday Launch Possible," *Florida Today*, August 27, 2006.

8. Roy LeCain, media briefing KSC, August 26, 2006.

9. Michael Cabbage, "Ernesto Likely to Strike Florida," *Orlando Sentinel*, August 28, 2006.

10. STS-115 Mission Control Center, "Status Report 1," September 9, 2006.

11. Steve MacLean, Conversation with the author, February 19, 2007.

12. Steve MacLean, "Sky-High Anxiety," *Toronto Sun*, October 11, 2006.

13. John Schwartz, "*Atlantis* Is in Good Shape as Its Crew Goes to Work," *New York Times*, September 11, 2006.

14. Steve MacLean, Conversation with the author, February 19, 2007.

15. STS-115 Mission Control Center, "Status Report 9," September 13, 2006.

16. "Astronauts Take Care of Business in the Sky," *Toronto Star*, September 14, 2006.

17. Tu Thanh Ha, "MacLean's Renovation Is Out of This World," *Globe and Mail*, September 14, 2006.

18. _____., "For Astronaut, Variety Is the Life of Space," *Globe and Mail*, October 17, 2006.

19. Associated Press, "Mystery Object Delays Space Shuttle Landing," *New York Times*, September 19, 2006.

20. Peter Rakobowchuk, "Astronauts Dodged a Bullet," *Toronto Star*, October 11, 2006.

21. STS-115 Mission Control Center, "Status Report 23," September 20, 2006.

Chapter 17: Build the Station, Build the Future

1.   STS-118 Mission Control Center, "Status Report 2," August 9, 2007.
2.   Traci Watson, "After 4 Years of Rehabilitation, *Endeavour* Returns," *USA Today*, August 8, 2007.
3.   Kenneth Chang, "Shuttle *Endeavour*, with *Challenger* Legacy, Takes Off," *New York Times*, August 9, 2007.
4.   Sidhartha Banerjee, "Astronaut and Family Try to Stay Grounded," *Toronto Star*, August 5, 2007.
5.   Marcia Dunn, "Teacher in Crew Helps Inspect for Damage," *Orlando Sentinel*, August 10, 2007.
6.   Tu Thanh Ha, "NASA Fears Ice Has Gouged Shuttle's Heat Shield," *Globe and Mail*, August 11, 2007.
7.   Dave Williams, Conversation with the author, June 16, 2008.
8.   Kathy Sullivan, "Walking in Space," *National Geographic Encyclopedia of Space* (Des Moines: National Geographic Books, 2004), 248.
9.   Rasha Madkour, "Ripped Glove Forces Early End to Spacewalk," Associated Press, August 15, 2007.
10. Dave Williams, Conversation with the author, June 16, 2008.
11. *Ibid.*
12. "Astronaut Answers Sask. Students' Questions," Canadian Press, August 20, 2007.
13. STS-118 Mission Control Center, "Status Report 11," August 13, 2007.
14. Marcia Dunn, "NASA Ponders Repairing Shuttle," Associated Press, August 13, 2007.
15. *Ibid.*
16. Liz Austin Peterson, "NASA Weighs Shuttle Repairs," Associated Press, August 14, 2007.
17. "NASA Decides No Shuttle Repairs Needed," Associated Press, August 16, 2007.
18. Marcia Dunn, "Shuttle Crew Prepares for Early Return," *Globe and Mail*, August 20, 2007.
19. Kenneth Chang, "NASA Moves Up Landing Because of Hurricane," *New York Times*, August 19, 2007.

20. STS-118 Mission Control Center, "Status Report 21," August 18, 2007.

21. Kenneth Chang, "Space Shuttle Lands Safely Despite Gouge," *New York Times*, August 21, 2007.

22. Dave Williams, Conversation with the author, June 16, 2008.

Chapter 18: The Forever Frontier

1. Liz Austin Peterson, "Williams' Spacewalk Shortened by Dean," *Globe and Mail*, August 18, 2007.

2. Linda K. Glover et al., *National Geographic Encyclopedia of Space* (Washington: National Geographic, 2005), 106.

3. NASA "STS-125: The Final Visit," June 16, 2008.

4. Drew Feustel, Conversation with the author at Johnson Space Center, August 8, 2007.

5. NASA "STS-125: The Final Visit."

6. CSA "Backgrounder," February 2, 2008.

7. *Ibid.*

8. Gary Kitmacher, ed., *Reference Guide to the International Space Station* (Burlington: Apogee, 2006), 3.

9. *Ibid.*

10. Lydia Dotto, *Canada in Space* (Toronto: Irwin, 1987), 143.

11. Canadian Space Agency, August, 2007.

12. *New York Times*, March 17, 2008.

13. Brian Laghi and Campbell Clark, "Ottawa Rejects Space Firm's Sale to U.S.," *Globe and Mail*, April 10, 2008.

14. NASA, *Spinoff* (Washington: Publications and Graphics Department, 2006), 51.

INDEX

CANADIANS IN SPACE